MICE & VOLES

MICE & VOLES

• JOHN FLOWERDEW •

with illustrations by
STEVEN KIRK

Whittet Books

First published 1993
Text © 1993 by John Flowerdew
Illustrations © 1993 by Steven Kirk
Whittet Books Ltd, 18 Anley Road, London W14 0BY

Design by Richard Kelly

The author and publishers are grateful to the following for permission to use
material in the graphs and tables in the book (page numbers where they appear in
brackets): P.B. Churcher and J.H. Lawton, *Predation by domestic cats in an
English village* (The Zoological Society of London) (p.19); R.J. Berry, *History in
the evolution of* Apodemus sylvaticus *at one edge of its range* (The Zoological
Society of London) (p. 32); R.S. Ostfeld, *The ecology of territoriality in small
mammals* (Elsevier) (p. 50); D.J. Baumgardner, S.E. Ward, D.A. Dewsbury,
Diurnal patterning of 8 activities in 14 species of muroid rodents (The
Psychomonic Society) (p. 54); L.J. Warner and G.T. Batt, *Some simple methods
for recording wild harvest mouse distribution and activity* (The Zoological
Society of London) (p. 57); R.J. Fuller and M.S. Warren, *Coppiced woodlands*
(NCC) (p. 63); I.R. Taylor, A. Dowell, T. Irving and G. Shaw, *The distribution
and abundance of the barn owl in south-west Scotland* (I.R. Taylor Scottish
Ornithologists Club) (p. 82); W.I. Montgomery, Peromyscus *and* Apodemus
(Texas Tech University Press) (p. 84); S.K. Alibhai and J.H.W. Gipps, *The
population dynamics of bank voles* (The Zoological Society of London) (p. 85);
R.C. Trout, *An ecological study of populations of wild harvest mice* (Blackwell
Scientific) (p. 86); Gurnell, Hicks and Whitbread, *Ecology and management of
coppice woodlands* (Chapman & Hall) (p. 64); D.W. Yalden, *Dietary separation
of owls in the Peak District* (Blackwell Scientific) (p. 94); G.B. Corbet and S.
Harris, *The handbook of British mammals* (Blackwell Scientific) (p. 30).

British Library Cataloguing in Publication Data
A catalogue record for this book is available from the British Library

ISBN 1 873580 08 8

Typeset by Litho Link Ltd, Welshpool, Powys, Wales
Printed and bound by Biddles of Guildford

Contents

Preface

In writing this book I hope to provide an easily accessible summary of the lives, ecology and behaviour of the British small mammals. I concentrate on the mice and voles but, for the sake of completeness, add necessary information on the shrews. It is inevitable that some important subjects will be omitted from a book of this size but further details and additional material can, hopefully, be found in the references listed in Further reading. I deliberately emphasize aspects of mouse and vole biology that I find particularly interesting, some of which may not be well known; I hope that the result will be informative as well as amusing!

The information I have drawn upon comes from a large number of sources and it is impossible to acknowledge all the people who have made contributions through their writings and research. I am, however, greatly indebted to Dr Carolyn King for keeping me up-to-date on the research she has been involved with in New Zealand and for drawing together the work which she and others have done on predation and on house mice; also, Dr Adrian Meehan's comprehensive review of house mouse (and rat) biology was a frequent companion while writing (their books are listed in Further reading). Dr Ian Montgomery has been a source of inspiration through his many research papers on wood mice and yellow-necked mice, Dr Donald Piggott kindly sent me the references to his work on damage to woodland seedlings, Rentokil Environmental Services at East Grinstead provided recent information on house mouse behaviour and ultrasonic repelling devices, and the books in Further reading by Dr Gordon Corbet and Professor Stephen Harris and by Drs Pat Morris and Paul Bright have been indispensable; Michael Woods kindly offered advice on setting up small mammal tables and mouse boxes. I must also acknowledge the many friends and colleagues I have met and talked to through the Mammal Society and in carrying out research; they are primarily responsible for stimulating and broadening my knowledge of small mammals over the years. Dr Paddy Sleeman has encouraged me throughout the writing of the book and the artist, Steven Kirk, has been a valuable source of ideas and discussion. Both Steven Kirk and Annabel Whittet have shown great understanding and patience while the book was in gestation.

Dr John Flowerdew

Town mouse, country mouse

Mice, voles and dormice, together with the shrews, are collectively known as small mammals; they are often seen scurrying away across a road or path (and you may have your own resident population); perhaps more often, one is found dead in the garden or dangling from the mouth of the cat that caused its demise. Although they are all secretive, they are some of the more commonly seen mammals, and most of them are extremely numerous and widespread. In mainland Britain we can find seven species: the **house mouse, wood mouse** (or long-tailed field mouse), **yellow-necked mouse, harvest mouse, common dormouse, short-tailed vole** (or field vole) and the **bank vole** – which, although they are not all very closely related, have many characteristics in common and belong to the same group that includes the rats, gerbils and hamsters, being part of the large mammalian order called the 'rodents' – the gnawing mammals. The small rodents discussed in this book seldom reach 50 g (2 oz) and so the water vole, edible dormouse and the rats are not covered; the common dormouse is the subject of a future book in this series. There are also three species of shrew, the common, pygmy and water, present on mainland Britain; these belong to the mammalian order called the 'insectivores' – the insect-eaters – and are sufficiently different in their biology to deserve separate treatment from the mice and voles. This book concentrates on the six smaller mice and voles but essential information on the identification and biology of shrews and common dormouse will also be found.

The British mice and voles are amongst the most abundant mammals in the country, it being debatable whether the **house mouse** or the **wood mouse** has the highest numbers. The **house mouse** (which is not, as its name suggests, purely found in houses) is one of the most successful agricultural, industrial and domestic pests in the world, as well as being an important laboratory animal in its domesticated form. Other species are not so numerous: the common dormouse is becoming rare and is protected by law, the yellow-necked mouse and harvest mouse have a restricted distribution and there are also a number of island forms of some species that exist in relatively low numbers. On the Orkney Islands and on Guernsey there are completely different 'common voles' which probably came from Continental Europe (see Island mice and voles).

What is a rodent?

The mice and voles found in Britain are only a small selection of the very successful rodents. This group of mammals is found all over the world and they possess many characteristic features. They have a pair of continuously growing front (incisor) teeth in each jaw which are made of tough enamel only on the outside surface. When the two pairs of incisors are moved against each other as the jaws work up and down, they are sharpened to form a chisel-like edge. In the burrowing mole rat (from Africa) the lower incisors can even move apart like forceps to help with digging. (Note that the rabbits and hares are grouped into a separate mammalian order, the Lagomorphs, and are easily distinguished from the rodents by the fact that they have two extra incisors in each jaw.) There are no eye (canine) teeth in rodents and instead there is a long gap in the jaw between the incisors and the cheek teeth (molars) called the 'diastema'. This gap allows the lips to be drawn in towards each other so that when the rodent is gnawing with the incisors to get at its food no stray pieces of material enter the mouth – as might be a problem when gnawing nuts or even lead pipes. The molars rarely exceed five in the upper jaw and four in the lower jaw. These cheek teeth may be cusped, like human teeth, as in many species that eat animal and vegetable material (omnivorous) and the few meat-eating (carnivorous) species, or form a flat grinding surface as in many grass- and vegetation-eating (herbivorous) species. The external appearance of rodents varies: there are jumping forms with long back legs such as the jumping mice, running forms with relatively short legs like the mice, voles and ground squirrels, burrowing forms like the gophers, tree-climbing forms like the squirrels, swimming forms like the beaver and coypu, and even gliding types like the flying squirrels. Some, like the porcupines, have spines over much of their body instead of hairs.

Although the ranges of rodent models reach extremes of over 30 kg (66 lb) with capybara and beaver (the latter extinct in Britain), the British species only go up to about 6 kg (13 lb) for the coypu (now exterminated) and the squirrels, which weigh up to 720 g (25 oz). Most of the British mice and voles weigh 20-30 g (about an oz) and are seldom more than 10 cm (4 in) long plus a tail.

The limbs of rodents are mostly used for a 'flat-footed' mode of locomotion, with the heel placed on the ground. However, the size of the feet and their adaptations are very varied, ranging from an elongated form in the two-footed jumping jerboas and spring hares and the shortened and clawing digging aids in the burrowing mole-rats, pocket gophers and tuco-tucos to the webbed feet of the beaver. Rodents usually have a rapid rate

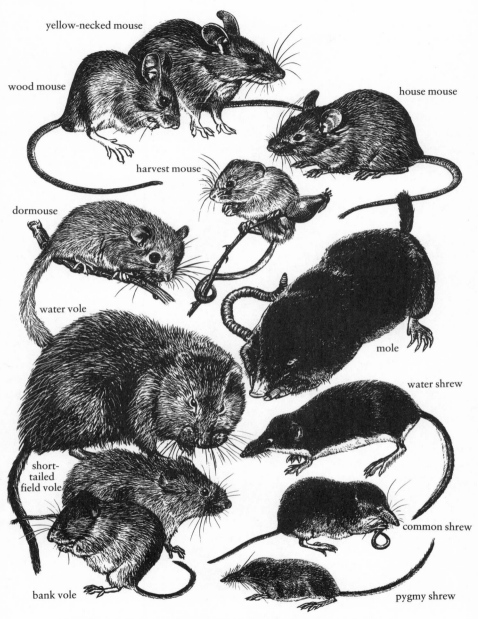

yellow-necked mouse

wood mouse

house mouse

harvest mouse

dormouse

water vole

mole

water shrew

short-
tailed
field vole

common shrew

bank vole

pygmy shrew

British mainland small mammals and some of their larger relatives.

of reproduction with several young in a litter and more than one litter each year, even if environmental conditions prevent reproduction in the winter.

The species discussed in this book have adapted to almost every habitat available and an enormous amount of knowledge has accumulated about them. Huge volumes are devoted to summarizing the biology of the **house mouse** and a substantial amount is known about the **wood mouse**, the **bank vole** and the **short-tailed vole**. The **common dormouse, harvest mouse** and **yellow-necked mouse** are moderately well studied but sometimes in much less detail: there is plenty of scope for new research!

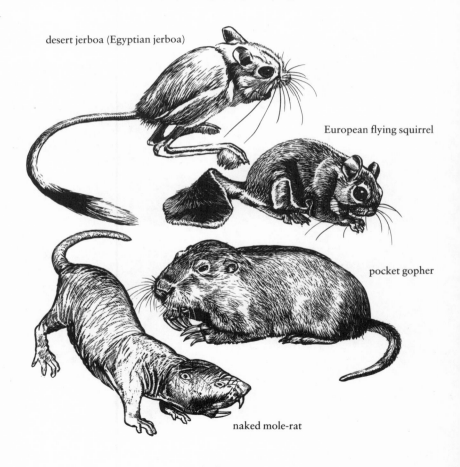

desert jerboa (Egyptian jerboa)

European flying squirrel

pocket gopher

naked mole-rat

What the cat brought in

Four mouse species, the common dormouse and two voles occur on mainland Britain and in addition there are three shrews which will be briefly described here (for information on shrews see p.122). Briefly, to tell the difference between a mouse, a vole and a shrew: mice have pointed noses, which do not extend beyond their front (incisor) teeth, prominent eyes and prominent ears and always have relatively long tails, at least three-quarters the length of the head and body. Voles, on the other hand, have rounded snouts with less prominent eyes than mice and ears which are partly or almost completely covered by fur. The tail length of voles is always less than three-quarters of the head and body length. The common dormouse, with a round snout, large eyes and a hairy tail, is unmistakable. The shrews have long, flexible, pointed snouts which project well beyond the front teeth and very small eyes and ears almost covered by fur.

The descriptions below outline the distribution, habitats and history of each species.

HOUSE MOUSE (*Mus domesticus*)

The **house mouse** is grey-brown above and usually lighter grey below; its head and body measure about 70-90 mm (3-4 in) and the tail is at least three-quarters the length of the head and body. The tail is more obviously scaly than in the other mice. The house mouse weighs 17-20 g (⅔ oz). You may find it difficult to distinguish small mice from young **wood mice** or **yellow-necked mice** because they also have grey fur. If you happen to have an accurate ruler with you, you could measure the hind feet: those of the **house mouse** are smaller. In adults these measure about 17-18 mm, whereas those of the **wood mouse** are 21-23 mm and **yellow-necked mice** 22-24 mm. Look also for the white belly fur, bulbous eyes and haired tail of the adult **wood mouse** and the additional yellow collar of the **yellow-necked mouse**.

The British species, *Mus domesticus*, is recorded throughout the country and occurs on most inhabited small islands. It is also found in Ireland and in western Europe; however, it is replaced by the very similar *Mus musculus* in Scandinavia and eastern Europe, a species which has lighter underfur and a shorter tail. The **house mouse**, or its close relatives, is found all over the world because of its association with man. It is

House mice have grey-brown fur and a scaly tail.

typically found in buildings in town or country, but may inhabit agricultural land, hedgerows and grassland. It commonly moves out of buildings in summer and autumn and back in winter.

House mice have been present in Britain since at least the Iron Age, 2,500-3,500 years ago.

WOOD MOUSE or LONG-TAILED FIELD MOUSE (*Apodemus sylvaticus*)

The **wood mouse** or **long-tailed field mouse** is well named because it is commonly found in woods and fields and has a tail which is almost as long as its head and body (80-100 mm, 3-4 in long). It is dark-brown above with some lighter brown and yellow along the flanks. The dark fur is clearly marked off from the lighter grey-white hair underneath and there is often a yellow chest spot of variable size which always has white or grey between it and the brown upper fur. The tail is lightly haired with dark skin above and light below. Adults and juveniles may be confused with the **yellow-necked mouse**; young ones with the **house mouse** because of their grey juvenile fur (see p.11).

Wood mice are found throughout Britain, Ireland and most of the islands (absent from Lundy, Isle of May, North Rona and the Isles of Scilly, except St Mary's and Tresco). They are very numerous and are found throughout Europe except northern Scandinavia and extend eastwards to the Altai and Himalayan mountains and south to northern Arabia and North Africa. The species has probably been present on mainland Britain since the end of the Pleistocene glaciations, at least 9,500 years ago, when it was still joined to the Continent. However, the origin of

Yellow-necked mouse or wood mouse?

The two common species of mouse found in woodland in southern Britain are the **yellow-necked mouse** *(Apodemus flavicollis) and the* **wood mouse**, *or* **long-tailed field mouse** *(A. sylvaticus). They are usually easy to tell apart because the yellow-necked mouse, as its name suggests, has a yellow-brown collar of hair across its neck and dipping down its chest between its front legs. The wood mouse has no such collar; and if it has any coloured fur at all it is usually confined to a small spot in the middle of the neck or a longer stripe running down the chest. However, just to make life more difficult, some wood mice occasionally have broader stripes of yellow-brown hair almost reaching the brown upper fur, making it*

Yellow-necked mice are more vigorous and one and a half times the size of wood mice.

difficult to tell the two species apart.

The other distinguishing characteristics of the yellow-necked mouse are that they are usually one and a half times as heavy as the wood mouse and even more frisky/noisy/likely to bite than a wood mouse when handled. The base of the tail in the male is particularly thick in the yellow-necked mouse, before becoming pinched in as it joins the body.

Other ways of telling them apart get rather technical: if you happen to have found a dead one, the depth of the front teeth (incisors) in the upper jaw may be measured across the unworn upper part; in the wood mouse they usually measure 1.1-1.3 mm and in the yellow-necked mouse they usually measure 1.45-1.65 mm. If the measurement is in between these ranges, because you have an old wood mouse or young yellow-necked mouse, the molars (cheek teeth) should be examined: if the cusps are well worn leaving a flat surface it is likely to be an old wood mouse whereas if the cheek teeth still have prominent pointed cusps showing little wear, it is likely to be a young yellow-necked mouse. This distinction is particularly useful for the identification of remains in owl pellets.

Wood mice usually come out to feed at night.

Mouse tails

The tail of the **house mouse** is relatively broad and strong in comparison with the other species of mice, so that they may be held (for a short time) by gripping the tail. Normally mice are held by the scruff of the neck and steadied by holding the tail. However, the **wood mouse** and **yellow-necked** **mouse** have a tail which is much less robust. If the mouse is held only by the tail then the skin is easily pulled off, allowing the animal to escape with bare vertebrae showing. This is obviously a good escape mechanism and the mice appear to come to no harm as the exposed vertebrae dry up and fall off before long. It is therefore important in handling these species that the animal is never held only by the tail in order to prevent damage and escape!

many of the island populations is more varied. Skeletal evidence suggests that many of the Scottish island populations, including those on the Hebrides, came (with the Vikings?) from Norway (see Island mice and voles). The populations on the Channel Islands probably came with man from the Continent.

Typical habitats are hedgerows, woodland, grassland and arable land including ploughed fields, but they will live almost anywhere, and will enter outbuildings in the winter, making nests in sheds and garages.

YELLOW-NECKED MOUSE (*Apodemus flavicollis*)
The **yellow-necked mouse** is very similar to the **wood mouse** in shape and colour except that it is larger in size (head and body: 90-130 mm, 4-5½ in) and has a small difference in its markings. The **yellow-necked mouse** is about 1.5 times the weight of a wood mouse (30 g, or more, compared with 20-25 g) and it has a yellow collar on the neck, sometimes extending down the chest. (The **wood mouse** has only white-grey hair on its chest and belly, with the possibility of a brown spot instead of the yellow collar.) No white interrupts the yellow collar from merging with the brown upper fur on either side so it forms a continuous band around the neck rather than a spot, and this is probably the easiest way of distinguishing it from the **wood mouse**. Young individuals also show the yellow collar amongst the grey juvenile fur. **Yellow-necked mice** are notoriously vigorous when held in the hand and often try to bite; they may also emit a high-pitched scream when held as will **wood mice**.

Yellow-necked mice are present in Britain only on the mainland in the south and south-east, in mid-Wales and the Welsh marches and in a few

Mouse squatters

Both **wood mice** and **yellow-necked mice** are agile climbers and they will often spend some time foraging in hedges and in trees. Because of these arboreal habits they may build nests in holes in trees above the ground and in bird's nest boxes; one nest has even been found with mice in it in the tool box of an old potato-digger left by a hedgerow in a field. When bird's nest boxes and dormouse boxes are cleared of debris one or more mice of either species can commonly be found. Nests will also be found in holes in trees, in garages and in lofts (see Mouse boxes in Habitats, burrows and nests).

Yellow-necked mice have a yellow-brown collar joining the brown upper fur on both sides.

isolated areas of south-west England; they are largely absent from the Midlands and the North, and also from Ireland. In Europe they extend further north than the **wood mouse** into Scandinavia but are more restricted elsewhere; they are generally a more mountainous species as they move down to southern Europe.

Mouse on the menu

The prey of about seventy cats in a village near Bedford were studied over a year, yielding over a thousand items, which were mainly mammals and birds. Most of the prey were caught in the summer, but the proportion of birds caught was highest in December and January; perhaps birds are easier to catch then and small mammals tend to stay underground in severe weather conditions. Birds also comprised over 50% of the prey in June – when many fledglings are around in gardens.

Wood mice, bank voles, short-tailed voles, common shrews and **pygmy shrews** comprised over 50% of the prey, with rabbits, other mammals and birds, house sparrows, blackbirds, song thrushes and robins making up the rest. The pie chart shows these results. The seasonal distribution of the two commonest prey, **wood mice** and house sparrows, showed that the mice were caught most often from September to December and again from March to May, whereas sparrows were caught mostly from June to September and again in January.

Cats particularly like catching wood mice, sparrows and voles.

Records of specimens from 5-6,000 years ago and later suggest that the species was formerly more widespread in Britain than it is today and that it was an early natural immigrant but early fossil evidence from the post-glacial period about 10,000 years ago is needed to confirm this.

Typical habitat in England and Wales is woodland and old hedges and there seems to be some connection with ancient (unchanged) deciduous woodland. They are only occasionally found in other habitats, but will enter houses and outbuildings even more commonly than **wood mice**, being notorious stored-apple eaters.

HARVEST MOUSE (*Micromys minutus*)

The **harvest mouse** is the smallest of the British rodents, weighing 5-10 g (⅓ oz); head and body measure 50-70 mm (2-3 in) long. It is russet-orange in colour and the adult has white underparts. Juveniles have grey-brown (sometimes sandy yellow) fur and adults appear darker orange-brown in the winter because of the greater density of long black-tipped guard hairs. The mouse has relatively small eyes and ears and a relatively rounded muzzle. However, the length of the prehensile tail (almost as long as the length of the head and body) firmly identifies it as belonging to the mouse family.

It is found from central Yorkshire southwards but seems to be absent

The not so deadly combine harvester

It was a common assumption up until the 1970s that the **harvest mouse** *was rare or decreasing, with a distribution much more southerly than a hundred years ago. The reason for this was supposed to be the steady increase in mechanical harvesting culminating in the combine harvester from the 1950s onwards. However, a Mammal Society survey in the mid-1970s showed that the distribution was really not very different from earlier times and that suitable*

habitats, including grassland, had thriving populations. Cereal fields and such food as wheat ears, for example, are not necessary for the animal's survival and combine harvesting does not necessarily kill animals in cereal fields. They are, however, probably less abundant than the **house mouse** *and* **wood mouse** *because of their Midland and southerly distribution and the lack of waste grassland, early coppice, reedbeds and other habitats with tall, dense vegetation.*

Harvest mice have bright orange-russet fur and a prehensile tail.

from north-east England, most of Wales and from Ireland. Outlying populations have been found south of Edinburgh and along the coast in north, mid- and south-west Wales; its presence on the Isle of Wight remains to be confirmed.

Habitats are commonly grassland environments with tall dense vegetation, including cereal fields, reedbeds, sedgefields, grassy hedgerows, ditches and bramble patches. Animals have been found in coppice and deciduous woodland but they are likely to have arrived there from other habitats.

Evidence from archaeological records suggests that the **harvest mouse** was introduced to Britain about 4,500-6,000 years ago or later from the Continent. This argument is based partly on the lack of fossil evidence before this time and records in deposits from Roman times onwards, and so should be treated with caution.

BANK VOLE (*Clethrionomys glareolus*)

The **bank vole** is chestnut-brown above and cream-white below with a tail (40-50 mm, 1½-2 in) just less than half the length of the head and body (90-110 mm, 3½-4½ in). It weighs 18-25 g, just less than 1 oz. It has the typical vole features of the ears just showing through the fur, eyes prominent but not as large as those of mice and a rounded face and muzzle. Like the mice the juveniles are grey-brown before they grow the bright adult coat at 4-6 weeks of age.

It is found throughout mainland Britain and on a number of islands as well as in south-west Ireland where it was first discovered in 1964. One of the island forms, the **Skomer vole** (*Clethrionomys glareolus skomerensis*), is slightly larger and has a very placid temperament so that it is easily handled (see Island mice and voles). The mainland species has been recorded from Roman times onward but earlier remains are of uncertain identity.

The **bank vole** requires good cover; it is commonly found in mature deciduous and coniferous woodland and in hedgerows, grasslands and young plantations.

THE SHORT-TAILED VOLE OR FIELD VOLE (*Microtus agrestis*)

The **short-tailed vole** or **field vole** is grey-brown above and creamy-grey on the underside. It weighs 20-40 g (¾ to 1½ oz) and the head and body length is 90-115 mm (3½-5 in). The tail (less than 40% of head and body length) is very much shorter than that of a mouse, and shorter than that of a **bank vole**; the ears are almost completely covered by fur, whereas those

Bank voles are common in woodland and hedgerows. They have chestnut-brown upper fur.

of the bank vole protrude somewhat through the fur. Their eyes are prominent but relatively smaller than those of the bank vole. There are two further clues to this species: it often squeaks when handled and it has a characteristic smell, something like 'musky cheese'; if you pick it up without gloves, you will be left in no doubt as to which species you've been handling, especially if the vole has left a wet patch on your jacket.

Short-tailed voles are found throughout mainland Britain and form one of the most important prey items for many birds and carnivorous mammals. They are absent from a number of western islands off Scotland, Shetland, the Isle of Man, Lundy and the Isles of Scilly. They are also absent from Ireland. On Orkney and the Channel Islands they are replaced by similar but slightly larger **Orkney** and **Guernsey voles** (*Microtus arvalis*) which have links with the Continent of Europe (see Island mice and voles). The species is distributed across northern and central Europe and in isolated spots further east. Remains of **short-tailed voles** have been found in Britain dating back to before the end of the last glaciation.

The habitat of the **short-tailed vole** is typically ungrazed grassland but they are also common in the early stages of forestry plantations where there is much grass in between the trees. They will also live in woodlands, hedgerows, dunes, scree and moorland wherever some grass growth is available.

COMMON DORMOUSE OR HAZEL DORMOUSE (*Muscardinus avellanarius*)
The **common dormouse** has very distinctive orange-brown fur and a bushy tail. These features, together with its rounded face and large eyes, make it easy to differentiate from the other small mammals. Length of head and body is 60-90 mm (2-3½ in) and it weighs 15-25 g (½-1 oz). The **edible** or **fat dormouse** (*Glis glis*) is present in Buckinghamshire and surrounding counties but will not cause any confusion because it is substantially larger (140 g, 5 oz) and greyish in colour, like a young grey squirrel. The **common dormouse** is found in England from Leicestershire and Suffolk southwards and in parts of south and mid-Wales as well as in a few localized areas of England further north. The main habitat is deciduous woodland with much secondary shrub growth such as coppice; they need trees and shrubs where they can move about and feed above ground. **Common dormice** specialize in eating flowers, fruit and nuts.

Short-tailed voles or field voles are important food items for kestrels and many other predators.

Common dormice specialize in eating nuts, fruits and flowers and they hibernate in winter.

They usually hibernate from October to April in nests on or near the ground. They probably colonized Britain after the last glaciation with the development of woodland about 9,000 years ago but definite remains go back only to 2,500-3,000 years ago.

It seems appropriate at this point to give a description of the three shrew species although they will not be considered in detail in the book. All three species may be found on mainland Britain and they will commonly form part of any small mammal study. Further details of their biology can be found in the sections on teeth and shrews.

COMMON SHREW (*Sorex araneus*)

The **common shrew**, like all the shrew species found in Britain, has a long mobile snout, short rounded ears and very small eyes. It is distinguished from the other shrews by having velvety dark brown fur on top with a greyish-white underside clearly separated from the upper fur; young ones have a lighter brown colour above. The tail in younger individuals is covered in hairs, some of which project from the end and it measures just over half the length of the head and body. The ears sometimes have white tufts of hair around the edge. The hind feet have fringes of stiff hair. Measurements for head and body are 50-80 mm (2-3½ in); tail = 24-44 mm (1-1¾ in); they weigh 8-13 g (¼-½ oz).

Common shrews are distributed throughout northern Europe and as far east as Siberia. They occur throughout mainland Britain but not in Ireland or in the Outer Hebrides, Shetlands, Orkneys, Isle of Man, some Inner Hebrides, the Channel Islands and the Scilly Isles. Common habitats are grassland, scrub, hedgerows and woodland. **Common shrews** have probably been present in Britain continuously since the end of the last glaciation 9,500-10,500 years ago.

PYGMY SHREW (*Sorex minutus*)

The **pygmy shrew** is, as its name implies, the smallest of the shrews in Britain, weighing only 2-6 grams (¹⁄₁₀-¼ oz). The upper fur is mid-brown, not as dark as that of the **common shrew**, and its demarcation from the lighter belly fur is less distinct. The tail is relatively thick and long, being almost 70% of the length of the head and body, and there are no long fringes of hair on the feet or tail. Head and body = 40-60 mm (1⅔-2½ in).

Pygmy shrews are widely distributed throughout the whole of Europe except for parts of the Mediterranean and they occur as far east as Siberia and south to the Himalayas. They are found all over mainland Britain and Ireland and on most small islands; they are absent from the

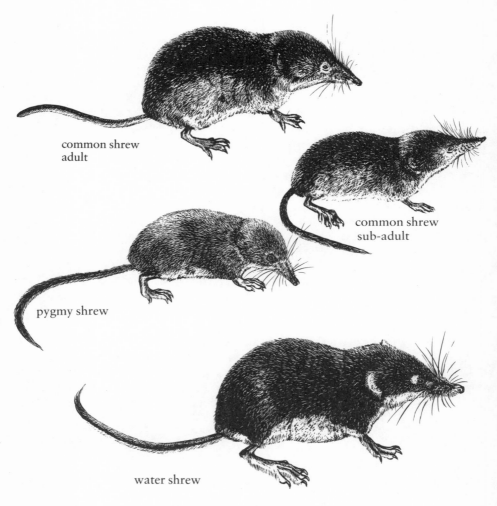

common shrew
adult

common shrew
sub-adult

pygmy shrew

water shrew

British mainland shrews.

Shetlands, Isles of Scilly and the Channel Islands, where a similar, closely related species is present. They are found in the same habitats as the **common shrew**, wherever there is ground cover, but they are usually less abundant; they are even found on moorland at high altitude. The **pygmy shrew** has probably been present for the same length of time as the **common shrew** since the end of the last glaciation.

WATER SHREW (*Neomys fodiens*)

This shrew is well adapted for swimming, with fringes of hair on the feet and tail. It is black above with belly fur of white, grey or black and is considerably larger than the other two shrew species, weighing 12-18 grams and sometimes more. Head and body = 67-96 mm (2¾-4 in).

The **water shrew** is distributed in Europe northwards to the Arctic circle, southwards to the northern Mediterranean and eastwards to Siberia. It is found throughout mainland Britain but only locally in the north of Scotland. It is absent from Ireland. It is commonly found near clean rivers, streams and water-cress beds but will be found in woodland, grassland and hedgerows, a long way from water. Its history is probably similar to that of the **common shrew.**

Reading the pellets

Owl pellets have been analysed from widely spread locations in the British Isles to give an idea of the distribution of some species that have restricted ranges. Pellets from barn owls confirm that the harvest mouse does not reach very far into Scotland and that the water shrew has a very localized distribution in the north.

barn owl

tawny owl

kestrel

Barn owl pellets are shiny black and those of the tawny owl are dull grey. Kestrel pellets are much smaller.

Island mice and voles

Small mammals living on islands around Britain are divided into two groups. There are species like the Orkney vole (*Microtus arvalis orcadensis*) on the Orkney islands and the Guernsey vole (*Microtus arvalis sarnius*) on Guernsey which are sub-species related to the European common vole, not found on mainland Britain. Then there is a group of variants of mainland species like the **wood mouse** and **bank vole**, some of which have been honoured by the status of a sub-species, others of which are so similar to the mainland species that they are not considered different enough to be given a sub-specific name.

The island sub-species and variants present a wide variety of size and to some extent colour, as well as skeletal characteristics, which are used to distinguish them from mainland forms. As a generalization, most of the island mice and voles are larger than their mainland counterparts, possibly because of the lack of ground predators; whereas on the mainland smaller animals have an advantage because they can escape into holes too small for the predator (see Predators and predation). But, in seeming contradiction, one Irish island has wood mice that are smaller than the mainland forms.

The European common vole (*Microtus arvalis*) has island forms on a number of Scottish islands and on Guernsey. They vary in size, colour and other characteristics as follows.

Variation in the European common vole and its island races.

	(subspecies of Microtus arvalis*)*			
	M.a. orcadensis	*M.a. sandayensis*	*M.a. sarnius*	*M.a. arvalis*
Range	Orkney: Mainland Rousay S. Ronaldsay	Orkney: Sanday Westray	Guernsey	W. Europe
Size	Large	Large	Large	Small
Back colour	Dark	Lighter	Lighter	Lighter
Max. skull length	30 mm	29 mm	28 mm	24 mm

The **Orkney vole** may have been introduced by man, as Neolithic settlements have been found dating from 5-6,000 years ago, or it may have been present since the time the islands were connected to the mainland, about 10,000 years ago. The vole was present on the Orkney island of Shapinsay in 1906 but is now apparently absent but *is* present on the mainland of Orkney, Rousay and S. Ronaldsay.

The **Guernsey vole** is similar to the Orkney vole (see table) but has a grey underside whereas the Orkney sub-species has a buff or creamy colour as well as grey, similar to the mainland species.

The **Skomer vole** (*Clethrionomys glareolus skomerensis*) – a sub-species of **bank vole** noted for its docility and the ease with which it is captured – is found only on the island of Skomer off South-West Wales and other sub-species of bank vole are found on Jersey, Mull and Raasay. As expected, all the island forms are larger than those on the mainland; the Skomer sub-species has a brighter chestnut coat colour and there are differences in the teeth and bones of the skull in all of them. It is thought that these island forms were introduced by man, probably with livestock or their food/bedding. There are many offshore islands with populations similar to the mainland.

Wood mice (*Apodemus sylvaticus*) show a lot of variation on islands but they are not different enough from the mainland form to be considered sub-species. They will interbreed with mainland forms and so are regarded as the same species. There are distinct forms on St Kilda, eleven of the Hebrides and on three of the Shetland islands, at least; those from the Hebridean islands of St Kilda and Rhum and Fair Isle (between the Orkneys and Shetlands) are particularly large. A soldier on St Kilda even reported that 'rats' had invaded and would kill all the ground-nesting birds but it was found that he had come from the Far East where the rats are small and he was, fortunately, mistaken! Tail lengths, body size, colour, teeth and skeletal features have all been studied and some fascinating conclusions drawn. Even their big feet have been suggested as adaptations to clinging on to windswept rocks!

Studies of skeletal variation (presence or absence of particular small bones and holes in bones of the skull) by Professor Sam Berry have shed light on the possible relationships between the island and mainland forms. He found that the closest mainland relatives of the Hebridean and Shetland wood mice were in Norway and that the mice were probably introduced to places of importance for Scandinavian voyagers, possibly the Vikings! The island types were not very closely related to those from the Scottish mainland and even the mice from St Kilda (way out to the

Map showing the suggested routes by which wood mice colonized Shetland and the Hebrides.

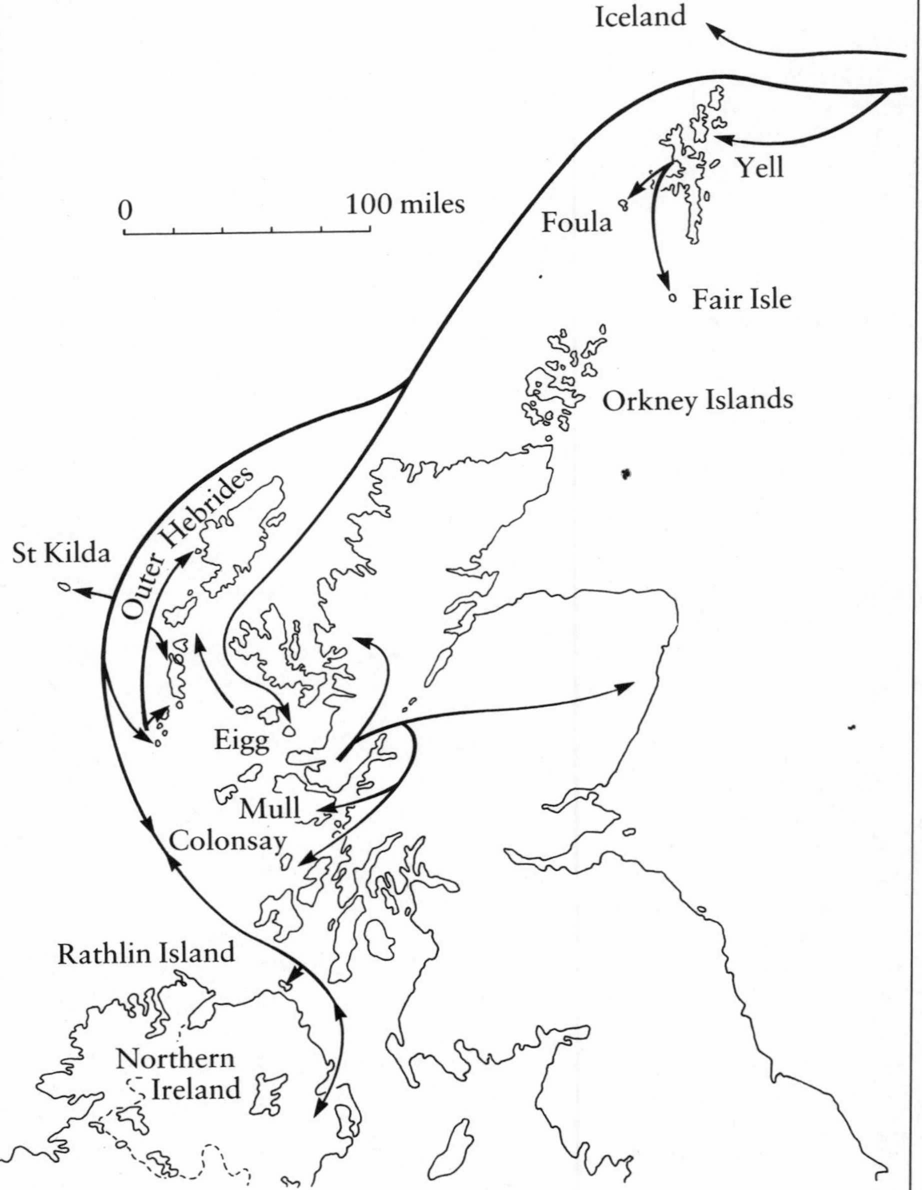

Iceland

Yell

Foula

Fair Isle

Orkney Islands

St Kilda

Outer Hebrides

Eigg

Mull

Colonsay

Rathlin Island

Northern Ireland

0   100 miles

Wood mice possibly came to many Scottish islands with the Vikings.

west of the outer Hebrides) and Iceland were more closely related to those from Norway. The skeletal studies, and historical analysis, suggest that some of the Inner Hebrides and Outer Hebrides were colonized first and that the other island forms were introduced from these cultural centres. Wood mice on the Shetland islands probably followed a similar pattern of primary introduction and then local colonization.

Research on the wood mice from the southern islands is more controversial. On Jersey, Guernsey and St Mary's, in the Scillies, the wood mice are similar to those on the mainland and may be remnants of pre-glacial times, although this has been disputed. On the small islands of Alderney, Sark and Herm the wood mice are similar to those on Guernsey and those on Tresco similar to those on St Mary's, suggesting that they were introduced from close-by sources.

Island **house mice** often show larger body size, slightly different fur

colour and skeletal variation. However, they are all considered to be the same species. On the island of Hirta in the St Kilda group, where they are now extinct, they used to be particularly large and were once considered to be a different species. The house mice became extinct on St Kilda after the evacuation of the human population in 1930; it appears that without humans they could not compete with the large wood mice on the island! Most inhabited small islands have populations of house mice and they have been studied extensively on Skokholm, off South-West Wales and on the Isle of May off the east coast of Scotland. Both populations live in cracks in the cliffs and in stone walls; they are both about 15% longer and heavier than mainland mice but they are genetically very different from each other, probably because the original colonizers were very different.

Orkney vole, showing uniformly light tail compared with short-tailed vole, more prominent ears, darker colouration; fur is thick, but less rough and shaggy compared with short-tailed vole.

Truth in a tooth

The skulls and teeth of small mammals provide vital clues with which to identify their owners. Bits of bone are frequently found in owl pellets – the regurgitated remains of meals which are left, often on the ground, near owl roosts – or in the castings of other birds of prey such as kestrels (see p.29). Small mammals often are attracted to discarded bottles, which then become a trap, because they cannot climb out on the slippery surface; the same thing happens with old drinks cans; both may become premature graves, and therefore may contain bones.

Once a skull has been found, it can be easily identified as belonging to the insectivore group of shrews (or the mole), or the rodent group of mice (or rats), voles and dormouse, by the shape of the skull and the nature of the cheek teeth. Shrew (and mole) skulls and jaws are elongated with no gap (diastema) between the front and cheek teeth. Mouse and vole skulls have a more blunt appearance with a definite gap between the front (incisor) teeth and the cheek teeth. Mice (and rats) have definite raised cusps (points) on the three cheek teeth in each jaw, with three or four separate roots fitting into sockets. The voles have a row of three jagged-edged flat cheek teeth in each jaw with a zig-zag pattern on their surface, all of which slot into one large socket, usually with no separate holes for individual roots. The **common dormouse** is easily distinguished because it has four cheek teeth in each upper jaw with faint, nearly parallel, cross-ridges on them with the tooth nearest to the front being only half as wide as the other three.

Which mouse or vole?
To tell the difference between the various mouse species the number of sockets for the roots of the first (front) upper cheek tooth must be examined. The **harvest mouse** has five, the **wood mouse** and **yellow-necked mouse** have four (see **yellow-necked mouse** or **wood mouse** section for further distinctions) and the **house mouse** has three. The Norway rat (*Rattus norvegicus*) and the black rat (*Rattus rattus*) also have five sockets but their skulls are much larger than that of the **harvest mouse**. For the voles, the two small species differ in that the **bank vole** has rounded edges to the cheek teeth when viewed from above and each tooth has two roots in older individuals, while the **short-tailed vole** has sharp, angular zig-zag edges to the cheek teeth and they always have an open pulp cavity

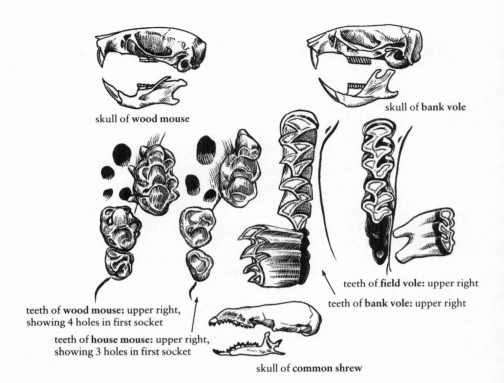

skull of **wood mouse**

skull of **bank vole**

teeth of **field vole**: upper right

teeth of **bank vole**: upper right

teeth of **wood mouse**: upper right, showing 4 holes in first socket

teeth of **house mouse**: upper right, showing 3 holes in first socket

skull of **common shrew**

Differences in skull shape, tooth types and the numbers of roots help identify small mammals.

underneath with no roots. In the water vole (*Arvicola terrestris*) the cheek teeth are similar to those of the **short-tailed vole** with open roots, and sharp edges giving a zig-zag pattern and the whole lower jaw is almost twice as big.

Shrew identification from skulls
The skulls of shrews found on mainland Britain all have teeth with a red coloured enamel which shows mainly at the tips. This contrasts with the white-toothed shrews (*Crocidura* species) which are found only in the Channel Islands and the Isles of Scilly, as well as on the Continent of Europe. The shrews have incisors in the upper and lower jaw which point forward, working like a pair of forceps, and the upper incisor always has two points (cusps); behind these on the upper jaw are a series of four or

The incisor teeth of rodents need to be continually worn down as they never stop growing.

five single-cusped (unicuspid) teeth which are used for identification purposes.

In the **common shrew** and the **pygmy shrew** there are five unicuspid teeth after the incisor on each side in the upper jaw. The **common shrew** has a larger skull, with the lower jaw, including the forward-pointing incisor, measuring about 12 mm. The third unicuspid tooth in the upper jaw is smaller than the second and the fifth uniscupid is very small indeed. In the **pygmy shrew** the skull is smaller, the lower jaw and incisor measure about 9 mm in length, and the third uniscupid is the same size as, or larger than, the second.

The **water shrew** has an even larger skull and a more concave forehead than the other two species and the lower jaw and incisor measure about 14 mm. There are only four uniscupid teeth in the upper jaw behind the upper incisor and the lower incisor has a distinctive, unridged, blade-like shape. The worn **common shrew** lower incisors may be confusing but if this is the case there will be little red pigment left behind on the other teeth.

Moulting and hair

Having a hairy coat is one of the fundamental characteristics of a mammal. The colour of the hair may be of use as camouflage, and the initial greyish colour of many small mammals may possibly help to identify a juvenile; but the main function of hair is to keep the animal warm. In order to do this the fur is changed after the first period of growth from juvenile to subadult and during adult life, sometimes seasonally, so that the small mammal can cope better with winter or summer environmental conditions.

Moulting and hair density

The way a mammal changes its fur is to moult so that areas of fur are lost and new fur grows in its place. This process has been studied in great detail in the **short-tailed vole**. The juvenile fur grows from specialized groups of cells in the skin called follicles which are arranged in groups of three and are fully developed over the skin soon after birth; the juvenile hair covers the body by 17 days of age and it is more grey in colour than the adult coat, as in most mice and voles, but not the **house mouse**. The second coat (present after the post-juvenile moult) begins to grow in a wave from the underside by 25 days, progressing over the flanks and upper parts to be complete by 45 days (called the 'sublateral' pattern of moult).

The coat consists of coarse overhairs, or guard hairs, and fine underfur. In the **short-tailed vole** the guard hairs and fine hairs on the back are coloured with two types of melanin pigments, one giving a black or brown colour and the other yellow or reddish-yellow. The pigmentation in the **short-tailed vole** occurs along the length of the hair except for a short region at the base. Hairs from the belly contain pigment close to the body but not further out. Thus the vole looks dark ash-grey/brown above and pale to dark-grey underneath.

In the winter the density of hairs is greater than in the summer and although the groups of hairs may be closer together because the animals are smaller in winter, there really are more hairs, mainly as a result of an increase in the number of fine hairs. There are also seasonal differences in the character of the guard hairs; in winter the wide region towards the end of the guard hair is shorter than in summer and the width at the base of the pigmented region is narrower in winter than in summer. Interestingly, these guard hairs are coarser in males than in females. These changes mean that in summer the coat is sparse with coarse hair and in winter the coat is dense with fine hair.

Thus in this seasonally moulting species the autumn moult gives rise to dense fur for winter conditions and the spring moult gives rise to less dense fur for summer conditions. Adults that have survived the winter moult in February; no hair growth occurs in December-January; few, if any, of these adults will survive to moult into the next winter. Juveniles develop a summer adult coat at the post-juvenile moult whether they are born early or late and then undergo an autumn moult sooner or later in preparation for the winter.

Moulting patterns

The way in which new hair grows often shows a characteristic pattern. The sublateral pattern is common in the post-juvenile moult of many species (e.g. **wood mouse, bank vole**) but adults often have a patchy 'diffuse' pattern of moulting.

In the **wood mouse** and the **yellow-necked mouse** the post-juvenile moult occurs at 5-7 weeks of age, lasting about two weeks, and the patchy diffuse moult of adults shows no regular seasonality although there are some indications that moulting is limited to the period August-January in some localities. The **yellow-necked mouse** post-juvenile moult leads to a coat that is not as bright as that found in older individuals, although the yellow collar coloration is always visible; it is thought that the dull colour is lost by a patchy moult as the mouse matures, possibly by about the 10th to 12th week of life.

In the **bank vole** the post-juvenile moult is from 4-6 weeks of age and leads to an adult moult, often of a diffuse nature. Peaks of adult moulting occur in spring and autumn. In the **house mouse** and **harvest mouse** the post-juvenile moult occurs at a similar time to the other mice and there is no seasonal moult in the former, which shows periodic waves of moulting fur moving over the body.

How to tell a mouse by its hair

With the aid of a microscope it is possible to look at the detailed structure of guard hairs (or nail-varnish casts of them) and see scale patterns on the surface. These patterns can be used to identify small mammals; the characteristic combinations of hair features found in the small rodents are too complicated to detail here but very good guides for identification are available (see Further Reading).

Coats of many colours

The natural colour (wild type) for small rodents will have evolved to be the best suited for camouflage, recognition, etc., for each species. However, different colour types of most species have been found. In the **house mouse** these are well understood: dark- or light-bellied forms, black (melanistic), albino, black-eyed white, pink-eyed with yellowish fur, spotted, leaden and cinnamon, and a general lightening of the colour may all be found. In **wood mice** black, white-spotted and silver-and pale-coated forms are recorded. There are no records of colour variants of **harvest mice** except for some individuals with white patches. In the **bank vole** colour variants are rarely recorded in the wild and in the **short-tailed vole**, black, black and tan, pink eye and normal coat, red eye and pale coat and piebald are found; black is very rare and albinos are rare.

Naked and black mice

*Completely hairless adult **house mice** and semi-hairless forms of **wood mice** have been found. In the **wood mice** the young develop normal hair and moult into an adult coat; however, the hair is lost rapidly from the head and by six months almost all the hair has gone, leaving them only with normal whiskers and tufts on the rump. Associated with this condition is lighter body weight, and presumably high mortality!*

*Melanistic or black forms of the **wood mouse** were found in 1975 in a sugar-beet field which had black peat soil near Ely. There is a little white fur around the mouth but otherwise they are completely black except for some pink skin on the*

*feet and inner ears. The black mice formed less than 5% of a large sample from this area. Coat colour is generally considered to be the result of selective predation and it is possible that there was some advantage in having black fur and presumably being more camouflaged on a black-soiled habitat. However, there is no evidence for this, especially as the proportion in the population was so low. It may be that some other less advantageous characteristics are also present in the black mice. It is well known that in **house mice** the coat colour genes are associated with other characteristics, some of which are not beneficial to survival — for example, obesity and sterility.*

Little breeders

Mice and voles are prolific breeders, producing 4 or more young in each litter and possibly 5 or 6 litters in a season. In addition, it is only a matter of weeks (5-6 weeks in species like the **house mouse** and **short-tailed vole**) before the young can produce litters of their own. The resulting capacity for population increase is remarkable and it is for this reason that some species like **house mice** and **short-tailed voles** become pests and numbers reach plague proportions, given suitable environmental conditions.

When living in a natural habitat, British mice and voles are seasonal breeders; the season usually starts in March or April and ends some time between October and December. The smaller **harvest mouse** doesn't start breeding until May and the **house mouse** in the wild usually stops by the end of September. However, house mice are an exception when living in association with man, when the great benefits mean they breed throughout the year. In woodland heavy crops of acorns or beech seeds will promote breeding throughout the winter or at least delay the end and help an earlier start in the spring, in the **wood mouse, yellow-necked mouse** and the **bank vole**. This also occurs in **house mice** living in beech forests in New Zealand where higher than usual numbers (known rather pompously as population irruptions) occur after autumns of abundant seed production. However, recent research suggests that it is the insects associated with the beech seeds that are important as food, rather than the beech seed itself. Where **yellow-necked mice** and **wood mice** occur together in Britain the **wood mice** usually start breeding some weeks later than the **yellow-necked mice. Bank voles** in woodland will often start breeding a few weeks after the mice and stop a week or so before them but there is much variation in the exact timing from one year to the next in all species. In **bank voles** and **wood mice** and **house mice** it is common for males to become sexually mature before the females at the start of the breeding season. However, laboratory studies of **house mice** indicate that males reach puberty at an older age than females.

It is likely that the length of light during the day (long periods of light in summer and short periods in winter) is a strong controlling factor in stimulating and inhibiting reproductive activity in most of the mice and voles. However, other factors such as nutrition, temperature and social factors can override the effect of day length. For example, if **wood mice** are reared in winter light conditions but with summer temperatures, they become sexually mature almost as quickly as mice reared in summer light and summer temperatures. In contrast, the **house mouse** indoors has breeding seasons which are not affected by day length; indeed, **house mice** are so adaptable that they will even breed in commercial deep-freeze warehouses, where they build nests inside carcases and feed on the frozen meat! When living outdoors, their reproduction is seasonal, and probably influenced by temperature and nutrition, as well as by population density.

Nests are obviously important places for eating or storing food and rearing young. They are discussed further in the sections on Habitats, burrows and nests and Mouse territories.

Litter sizes (see table, p.44) vary with age and with conditions such as availability of food and weather. Younger females are likely to have smaller litters than older ones and early litters are likely to be larger than later ones, so the average litter size found in the population of many of the seasonally breeding species changes from one month to the next, often reaching a peak in the middle and getting smaller as the season goes on and young litters of the year enter the breeding population. **House mice** usually have smaller litters in buildings than outside in ricks or in fields; this is thought to be the result of better feeding conditions outside, but may also be due to lower population densities (less aggravation?) and the fact that the mice breed at an earlier age indoors. In **wood mice** and **bank voles**, which continue breeding into the winter, the litter size is smaller in autumn/winter than in spring and summer.

The length of pregnancy (see table, p.44) is about 20 days in many of the mice and voles but gestation may be extended if there are young being suckled. In **wood mice** this is equivalent to an extra 1.3 days of pregnancy for every extra pup being suckled. This is called 'embryonic diapause': the growth of the fertilized egg is retarded and its implantation (usually in days 4-5 of pregnancy) in the uterus delayed. Successive litters may be produced throughout the breeding season with the mice being capable of producing 5 or more, except the **harvest mouse** which has a recorded maximum of 3. It is common for female mice and voles to become receptive to males immediately after they have given birth so that another pregnancy usually follows (lactation, which is a dubious contraceptive in humans, is obviously not one at all for mice).

Growth is rapid and young are usually weaned within about 2-3 weeks of birth, depending on the species. The 'pups' are born naked and blind and weigh about a gram, or less, in the smaller **harvest mouse**; between 1 and 2 grams in the **wood mouse** and closer to 2 grams in the **yellow-necked mouse** and the **short-tailed** and **bank voles**. At this early age they can make weak staggering movements to find the right place in the nest to suckle the female.

In the **wood mouse** the pink skin on the back and head soon darkens to produce a grey-brown velvety fur at about 6 days with whiteish fur on the underparts soon after. Teeth are present at about 13 days and the feet and tail darken on top after 14 days; the eyes and ears are open by about 16 days but precise timings vary with season and probably with litter size and feeding conditions. In the **yellow-necked mouse** the yellow collar is visible in the grey juvenile fur from about 2 weeks of age.

Litters of small mammals huddle together to keep warm in the nest.

Wood mouse pups soon grow fur but the eyes and ears develop later.

In the **bank vole** the bare skin in the new-born pup darkens by the 4th day after birth and the velvety-chestnut browny-grey juvenile fur starts to appear by day 8, covering the upper body fully by day 12; the white underfur appears just after the upper fur and the eyes and ears are open by day 12. The **short-tailed vole** probably develops in a similar sequence.

In the **harvest mouse** the young develop very quickly; short grey-brown hairs appear on their back by day 5 while the underside is still bare and scaly; white belly fur appears by day 7 and the eyes and ears open by day

9. They are likely to disperse from the nest by 16 days of age.

In the **house mouse** dark pigment appears in the skin of the bright pink pups at 5-7 days of age and hair is half grown by days 8-10. The eyes open at about 14 days and the teeth have erupted by day 14; weaning starts soon after.

Table of litter sizes and reproductive data in British mice and voles. (Note that mean litter sizes vary from one habitat/time to another)

Species	Litter size		Gestation length days	Lactation days	Pups eyes open
	range	mean			
House mouse	5-8	5.2-7.7	19-20+	18-20	14
Wood mouse	2-11	4.5-6.5	19-20+	18-22	16
Yellow-necked mouse	2-11	5.0-6.8	25	20	?
Harvest mouse	2-8	5.4	17-19	11-14	9
Bank vole	1-7	3.0-4.5	18-20	17-18	12
Short-tailed vole	2-7	3.0-6.7*	18-20	14-28	?

*from dissection and counts of embryos

Stranger on the block

*The **house mouse**'s capacity to reproduce has, for obvious reasons, been the subject of numerous laboratory studies, many of which involve the influence of chemical signals passed from one animal of the same species to another (sometimes called 'pheromones' or simply 'olfactory cues'). These substances are passed in the urine or possibly in secretions from specialized scent glands and may have an effect on the behaviour or physiology of the recipient (see also the section on smells, squeaks and behaviour).*

*In laboratory **house mice** it was found that a newly mated female would lose the fertilized eggs at an early stage of development if she was exposed during the first four days after mating to a 'strange' male (i.e. not the male she mated with). This 'pregnancy block' was called 'the Bruce Effect' after the researcher who discovered it and it is the result of a hormonal deficiency leading to a failure of the early embryos to implant in the uterus. The female will then return to a receptive condition, ready to mate again, and if the same or* another male is available (and ready to mate) she will become pregnant, giving birth after the usual length of time. Certain proteins in the male's urine have been found to be responsible for the pregnancy block. This type of effect on pregnancy has also been found in the **bank vole** and the **short-tailed vole**, but its occurrence in the wild is still to be confirmed.

Other physiological effects resulting from the male's influence on the female have been found in laboratory studies of **house mice**; again, it seems to be chemicals in the urine that have such an important effect. The induction of oestrus (physiological preparation of the female reproductive tract for mating and conception) was brought on when a male was introduced to a female previously housed with other females. If the male was introduced to several females who had been living together then they would all come into oestrus at the same time. The introduction of a male will also accelerate the onset of puberty in juvenile females previously housed without an adult male. Females grouped together may affect each other's physiology either by inhibiting puberty and oestrus or inducing pseudopregnancy (symptoms of pregnancy but without a fertile mating or a developing embryo), depending on who did the experiment. We don't know whether other mice and voles work the same way, but if groups of females ever occurred naturally it might be possible.

It could be advantageous for a female to breed as soon as possible in case she does not breed at all; in this way, she may be able to produce more litters in the breeding season than otherwise; females in groups will inhibit each other's reproductive potential. In the wild some females (usually the dominant ones) are likely to keep in breeding condition while the subordinate or younger females will be suppressed. Under these conditions (presumably at high density) it may be advantageous for the lower-ranking females not to breed if the dominant ones are likely to kill their litters .

Hanging on for dear life

Small rodents usually have six or ten nipples, two under the armpits and 4-8 arranged in pairs down the abdomen – enough to feed all the pups in a typical litter, and a few more. When the young are all feeding in the nest, the female may be alarmed by a predator or even another individual of her own species. In these circumstances one might imagine that there would be a conflict of interests for the female,

Pups must hold tight when mum is disturbed!

whether she should escape herself or remain to defend her progeny. However, there is an easy solution which helps both mother and young to avoid the danger. This is a phenomenon similar to that of 'caravanning' which has been observed in shrews. It is easily seen in captive female wood mice and bank voles. The pups hold on tightly to the nipple and the mother rushes off; she presents a rather large and ungainly appearance but usually all the young can be removed from the nest in one go – a lot more efficient than retrieving each pup by the scruff of the neck!

In the shrews, the phenomenon called 'caravanning' is more deliberate, as the young grasp the tail or the rump of the individual in front with the jaws; one description notes that the mother had six young in tow.

Male or female?

Sexing small mammals is probably easiest in the mice, slightly more difficult with voles and can be very difficult with shrews (see Shrews). Males and females are easiest to distinguish when you have both to compare them and when they are sexually mature. You can't always arrange this, however, and it is easy to do, with practice, on any individual except the very young.

The key to sex determination is the distance from the penis (male) or urinary papilla/clitoris (female) to the anus. (These distances are all pretty small, so make sure you've an accurate rule.) This distance is much greater in the male than in the female. If the male has well developed testes, then the penis-anus distance, for example of, the **wood mouse** may well be greater than 10 mm, whereas the clitoris-anus distance for even a mature

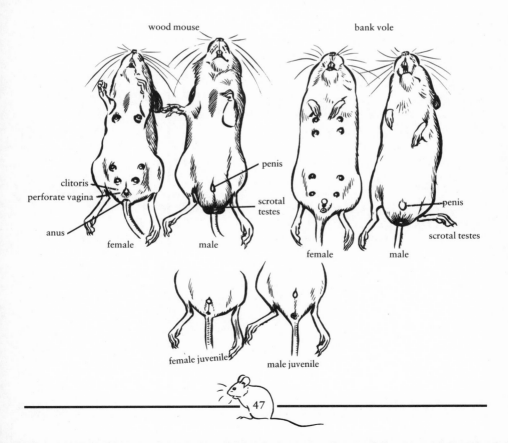

female is unlikely to be more than 5 mm. The opening of the vagina in females is found just below the clitoris in females which are in breeding condition and ready to mate or in the later stages of pregnancy. However, the vagina is not open in mid-pregnancy or during the non-breeding season or in juveniles and it may be indicated solely by a scaly covering of the perineal skin.

Females which are suckling young have obvious, well developed nipples – three-five on the left and three-five on the right. Male **wood mice**, **yellow-necked mice** and **house mice** in breeding condition have well developed testes which are obvious in the scrotal sac which may lack much hair at the tail end. However, in the **bank vole**, **short-tailed vole** and **harvest mouse** the testes are not as prominent and they may be seen only as swellings on either side of the lower abdomen. In **harvest mice** the breeding males have a penis which is thicker than that of young non-breeding males and the clitoris in the female, so making the sex-determination rather tricky.

Vaginal plugs

Female mice and voles that have been recently mated usually have a plug of coagulated semen left in the vagina called a 'copulation plug' or 'vaginal plug'. This helps in the retention of the sperm making conception more likely and may help prevent further mating, although it is likely that a male may be able to remove it. In **house mice** *the plug remains in position for 18-24 hours whereas in the* **harvest**

mouse *the female loses it quickly. In female* **short-tailed voles** *and* **bank voles** *copulatory plugs may be removed by further mating or remain and become fused with new plugs; the male may even eat a plug that has become dislodged! This means that young in the litter may be by different fathers; genetic studies have proved this in North American deer mice.*

Mouse and vole territories

Home ranges and territories

The use of space by small mammals is usually described in two dimensions although it obviously occurs in three dimensions. This is particularly true of the **harvest mouse** which makes use of tall plants as well as underground burrows, the **house mouse**, which may live in a stack of stored food or similar three-dimensional habitats and of the **yellow-necked mouse**, which climbs high into trees and uses runways along living and fallen branches particularly in autumn and winter. In the other species the assumption that they live on a flat surface is not far from the truth and radio-tracking of **wood mice** has shown that, apart from entering burrows and using nests, they do inhabit a more or less flat space.

When mice and voles move about looking for food and/or a mate they encounter other individuals of the same species or their smells. We know that when **bank voles** move about they deposit trails or drops of urine, and probably all the other small mammals do too. **House mice** literally cover their range with urine drops (see Smells, squeaks and behaviour): a new object is immediately obvious because it does not smell! This urine marking informs other members of the species who has been there; the reactions to neighbours and strangers may vary with season, the type of social system shown by the species, and the status of the home individual, adult or juvenile, dominant or submissive, receptive or antagonistic, courting or simply looking for food.

All small mammals have a home range – an area which they move over in their day-to-day activities – but the way they interact with other members of their species determines what sort of home range it is (see also Numbers). The movement patterns shown by the British small mammals can be classfied into three types: communal home ranges with a social hierarchy, individual home ranges which overlap those of others and individual territories which are mutually exclusive and possibly defended areas (this means that the reaction of a neighbour is to avoid the territory holder, even if overt aggression in defence of the area is uncommon). Territories may be shared, however, as in **house mice** (see Numbers).

In the wild-living species the use of space has been studied from a theoretical standpoint. One school of thought suggests that the spatial distribution of food, its renewal rate and its abundance determines whether females are territorial, while it is the pattern of availability of

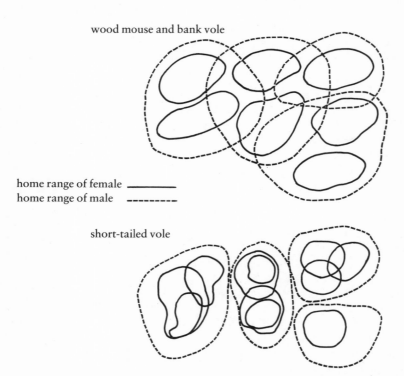

wood mouse and bank vole

home range of female _____
home range of male   - - - - - - - -

short-tailed vole

The use of space in mice and voles. In the short-tailed vole the female home ranges overlap but breeding males are territorial. In bank voles and wood mice breeding females are territorial and male ranges overlap.

females in space and time that determines whether males are territorial. Put simply, this means that if a small mammal female feeds on sparsely distributed seeds, fruits and herbs then it is beneficial for her to be territorial against other females and defend as much food as she can but allow males in for mating and feeding/exploration. Under the same circumstances it is beneficial for the male to be non-territorial and visit as many females as he can. Alternatively, if the female feeds on grasses, sedges and rushes which are abundant and evenly distributed as well as being rapidly renewed then she doesn't need to waste energy on being territorial and several females may overlap in their home ranges. In this system the males benefit by being territorial against other males and gathering a number of females within their territory.

The theory seems to stand up to the test: **bank voles** which are seed, fruit and herb eaters, have territorial females in the breeding season to protect as much as possible the sparsely available food; the males have overlapping home ranges and possibly interact through a dominance hierarchy. **Short-tailed voles**, which are grass eaters, have non-territorial females (one or more) living within the territory of a male. **Wood mice** behave like **bank voles** and sometimes two females appear to share the same territory. Not enough is known for certain about the social behaviour of **yellow-necked mice** and **harvest mice** but the latter move in overlapping home ranges, especially the males, and the females may be aggressive to each other, possibly defending breeding nests. It is possible that the male **short-tailed vole** might start the breeding season with a single female and then allow his female offspring to join the mother within his territory; this seems to happen in some North American voles.

The actual size of the home range or territory seems to vary with season and habitat. **House mice** in grain stores will move over much smaller areas than those in wheat fields and **wood mice** and **bank voles** will be non-territorial and move over smaller areas in winter when they are living in woodland, because there is plenty of food and they are not breeding. **Wood mice** move over much larger areas in agricultural land than in woodland and consequently live at much higher densities in woodland.

When the home range areas are calculated they differ according to the type of data collection (traps, radio-tracking etc.) and the method of calculation.

The size of a small mammal's home range will increase as it grows to maturity and is likely to increase under poor feeding conditions and decrease under good feeding conditions. Breeding males usually have larger home ranges than non-breeding males. **Wood mice, bank voles** and **short-tailed voles** often stay in the same area for long periods of time but they will move to new areas to exploit new food supplies and this particularly applies to the **wood mouse**. There is evidence for movements out of woodland in spring and back again in the autumn; in arable land home ranges in summer are probably larger than the larger ranges in woodland. In arable land movements of about 1 km have been recorded in **wood mice** moving to a field edge with particularly tasty weed seeds in it!

The shapes and sizes of **wood mouse** home ranges in woodland and sand dune habitats show dramatic differences. In woodland the ranges are irregular and relatively small in comparison with sand dune ranges which are much larger (males 36,499 m^2 and females 15,826 m^2 on average in the breeding season, probably because they are lacking much seed food) and usually circular in shape. There are presumably so few objects that might influence the boundary of the sand dune range that the mouse simply uses the most economical area, which of course is circular!

Movements away from the home range are probably to find a 'better' habitat for feeding or for mating. Female **wood mice** disperse more as the density of the population increases and as the food supply decreases. Male **wood mice** probably only move away in winter if they are attracted by particular food supplies and in summer if they are attracted by better mating opportunities.

Activity

Mice and voles tend to move out of their nests at particular times of the day. It is usual to classify **house mice, wood mice** and **yellow-necked mice** as nocturnal, only coming above ground during the night, and **bank voles** and **short-tailed voles** as diurnal, moving during the day and the night. **Harvest mice** are intermediate in being nocturnal in winter and more diurnal in the short days of summer.

'Activity' is perhaps not the correct word for what the **house mouse** does; the day is dominated by sleep, and a lot of time is spent motionless and in moving around (see table). Feeding is a major night-time activity but they will feed in the day if this is the only time when food is available.

Ranges of average values of estimated home range size from various habitats

Species	Season	Home range size (m^2)	
		Male	Female
Wood mouse	summer	1,284-13,063	809-4,009
	winter	299-1,294	242-1,151
Yellow-necked mouse	summer & winter	2,238	1,917
Harvest mouse	summer & winter	400	350
Bank vole	summer	929-1,398	271-953
	winter	380-1,209	261-1,067
Short-tailed vole	summer	600	200-300

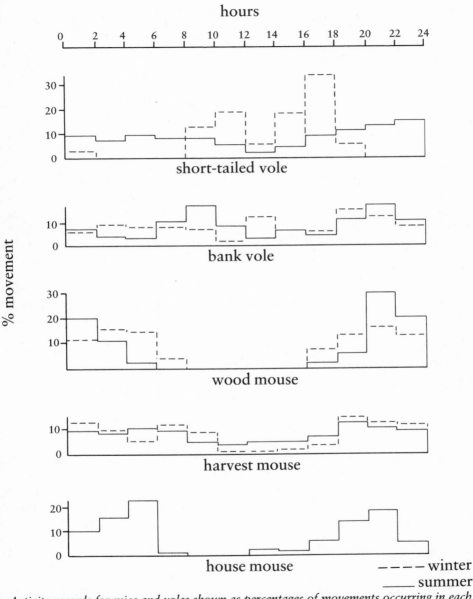

hours

% movement

short-tailed vole

bank vole

wood mouse

harvest mouse

house mouse

– – – – winter
———— summer

Activity records for mice and voles shown as percentages of movements occurring in each two-hour period of the day.

53

Activity budget for the house mouse in 8 hours dark, 16 hours light.

	Locomotion	Groom	Eat/ Drink	Sleep	Inactive	Other
Dark %	30	19	15	9	23	4
Light %	7	16	5	47	22	1
Total	15	17	8	34	23	2

Average time spent in various activities (shown as % of day)

The typical activity patterns of small mammals are shown on p.53 although most species are vary variable and can alter their activity greatly under some circumstances (e.g. disturbance, predation, social structure). Note that lactating **wood mouse** females may come out in the middle of the day during the summer, presumably to feed, and that movements of **house mice** and **wood mice** may be restricted on moonlit nights so that predators like owls will not have as much chance to see them. **Short-tailed voles** tend to be more diurnal in winter, probably trying to avoid the cold night temperatures. However, it has also been found that there is usually a two to four hour cycle associated with feeding: they come out to feed, fill their stomachs and then rest while the food is digested. This sequence of events is superimposed on the overall activity rhythm of the animal. **Short-tailed voles** show more daytime movements as they get larger (and probably more dominant), forcing the smaller animals to move about more in the night and dominant **wood mice** restrict the movements of subordinate individuals.

Mice and voles spend much of their time in one or more nests, scattered throughout their home range. Harvest mice even build special breeding nests to live in with the young. The time spent in the nest sleeping (or lactating for females) makes up a large part of the daily routine for wild small mammals. In a **bank vole** studied in captivity (which may or may not be typical of the wild) sleep made up 43% of its time, rest in nest 19%, alert periods 10%, grooming 13% and feeding/exploration 15%. Activity above ground occupied only 35% of the time! Nests of most species are commonly shared in winter to conserve energy but this is less common in the breeding season. In a radio-tracking study of **wood mice** in the breeding season the mice changed nests frequently and seldom shared nests, whereas in the non-breeding season communal nesting was common. This was corroborated by the fact that the radio-collars were chewed (presumably by other mice) in the non-breeding season but no sign of chewing was found in the breeding season.

Habitats, burrows and nests

Small mammals are secretive and difficult to see because they live underground or under the cover of vegetation for much of their lives. In addition, species like the **wood mouse** and **yellow-necked mouse** usually remain underground for most of the day and only venture out above ground at night (see Mouse territories). We know little about tunnel systems of species like the **wood mouse** and **bank vole**; at the moment we assume that each individual has its own burrow system and nest, although cases are known of **wood mice** and **bank voles** sharing nests with others of their own species in winter to help keep warm. Most of the species appear to lead solitary lives in the breeding season, meeting others only at preferred feeding sites and for mating. Females will, of course, be with young more of the time.

The **harvest mouse** lives in tall vegetation, especially grasses. In the summer most activity takes place above ground in the stems and leaves of plants in hedgerows, cereal fields, roadside verges and embankments, marshes, sand dunes and almost anywhere where tall grasses can be found. Its nest is made of woven grass leaves and may be 20 cm (8 in) above ground in a tussock or 150-200 cm (5-7 ft) in a bamboo clump. The nest may be a flimsy temporary shelter for one animal, about 4 cm (1½ in) in diameter, or an intricate breeding nest, measuring 6-10 cm (2½-4 in), or larger. The nest is supported by leaves between the stems and grass leaves are pulled inside and shredded with the incisor teeth to form many parallel strips before being interlaced with other leaf strips from the surrounding stems. The shredded leaves are still intact at the end near the stalk and so provide a firm foundation supporting the nest in the grass. The outside of the nest is green as here the grass leaves are still living. Inside fine grass strips are woven together to form a lining and other soft material such as feathers or thistledown may be used as well.

Breeding nests (used for one litter only) are usually found in tall grass stems and are made by the female in late pregnancy. Despite frequent visits in and out there are no obvious holes; this is because as the female squeezes past the shredded grass at one of several definite entry/exit points the elasticity of the strips closes the hole again. Non-breeding nests, used

55

Harvest mouse nests are supported by stems and have growing leaves weaved through them. (Breeding nest is on left; non-breeding or winter nest is on right.)

for eating, storing food and sleeping, are built between tall stems in the summer (presumably used by males) but in the winter, when no breeding occurs anyway, they are found in the bases of grass tussocks, in ricks and bales and even on the ground under suitable cover, as much more time is spent on the ground or in the burrows of other small mammals. The nests contain only weakly woven strips of grass, if they are woven at all.

Short-tailed voles are also grassland dwellers, living in ungrazed grassland, forest rides, road verges, young forest plantations with a good growth of grass and even in woodland and sedge/reed areas – wherever the grassy patches can support their feeding and living requirements. They make nests of about 10 cm (4 in) in diameter in the bases of grass tussocks, underground in a tunnel system or even under sheets of corrugated iron or asbestos (lifting these quickly is a good way to find voles and shrews). The nests are usually made of cut and shredded grass leaves and are therefore dried and brown when discovered. They have well formed entrances which lead out of the surface runway system. The runways are marked by

Are there harvest mice about?

If you suspect that harvest mice are present in an area but cannot confirm it by trapping or finding nests there is a simple method which can be used to test if they are present. All that needs to be done is to get hold of a few old tennis balls (they can be protected with water-proofing solution if you intend leaving them out for long) and make a single 15 mm hole in them with a cork borer. Each ball is held by a wire clip which is stapled to a wooden stake (rough wood so that the harvest mice can climb up it) and the stake placed in the ground so that the ball is about 30-50 cm (12-20 in) above the ground. Place a little budgerigar seed in each ball and leave the hole pointing to the side. You can position the stakes at sites where you suspect there may be harvest mice, amongst tall vegetation and grasses in hedgerows or reedbeds, etc. If there are many harvest mice present and they are foraging near the balls they should show signs of having eaten the seed within two days; if, however, they are at a low density they should have found the free meal within about 5 days.

Tennis balls baited with budgerigar seed are used to test for signs of harvest mouse feeding.

Nests of short-tailed voles are commonly found under metal sheets in grassland. Bank vole nests can be found in woodland.

cut grass stems and leaves, which may be dead and withered or still alive, and they weave their way through the grass tussocks and other vegetation, coming to the surface periodically and sometimes disappearing into the ground. The floor of the runway is commonly littered with small piles of cut grass stems and leaves and little 'hay piles' of similar material may be found at feeding sites, having a criss-cross jumbled appearance and often being supported by the surrounding tussocks. Faeces are also deposited, usually in groups at latrine sites throughout the runways.

Bank voles are commonly found in habitats with good ground cover, preferring dense stands of bracken and bramble in woodlands and similar habitats in hedgerows; thick stands of sedge and reed are especially favourable if not too wet. When they occur in grassland they are more common in scrubby areas where there is a varied vegetation of herbs and shrubs. Tunnel systems, often a few centimetres below ground, are excavated or possibly 'borrowed' from other small mammals and the nests are usually found underground; the tunnels commonly have food stores attached and lead to runways through the vegetation at ground level but it is difficult to tell these from those made by other species such as **wood mice**. The nests are made from material such as dead leaves, grass, moss and feathers and may be inside hollow tree trunks as well as underneath artificial cover such as corrugated iron or asbestos.

Wood mice are commonly found in almost every habitat from ploughed fields to ancient woodland. In woodland their nests are nearly always underground, often under a tree root system, or in a hole in a tree. However, they are also found in birds' nest boxes, and even dormouse boxes, as well as almost anything which will provide suitable cover such as agricultural machinery tool-boxes, garage floors and even old boots! The nests are made from any material that is handy, ranging from dead leaves, grass and moss to straw, shredded polythene sheeting and newspaper. They are of variable size and shape, often made to fit the containing structure, and they do not usually have a definite entrance; in some cases the ground nearby may be scattered with faecal pellets and in

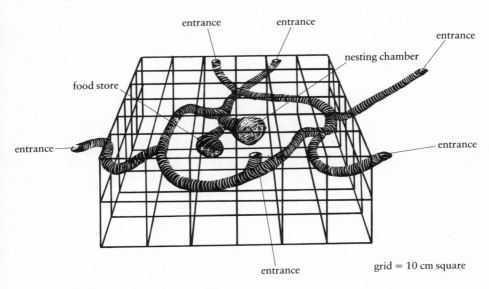

Wood mouse burrow systems in woodland have food stores and escape tunnels with the nest often under a tree trunk.

tunnel systems the remains of seed and insect foods may be found. The nests are not usually found under corrugated sheet metal or asbestos. **Wood mice** in stubble fields leave well worn paths to scattered holes in the ground and food stores in such habitats have been found to contain a few hundred grams of grain.

Tunnel systems have been excavated in woodland showing a series of entrances often leading to a ring tunnel circling the nest chamber and food store (see diagram); the tunnels may link with deeper mole runs and so provide a further means of escape from predators. Tunnels often run down a slope and may be quite extensive in suitable soil. New tunnels are often left with a pile of freshly dug earth outside. Characteristic of the **wood mouse** is the obstruction of the tunnel entrances with bulky material such as small stones, clods of earth or twigs which hide the hole; they may also block tunnels containing stored food so that live traps may be obstructed in this way with any combination of earth, stones and leaves.

Yellow-necked mice have similar nesting habits to **wood mice** but are much less likely to be found outside woodland. Both species will use runways through ground vegetation which are shared with other small mammals, and move on low-level branches. **Yellow-necked mice** are generally considered to be more arboreal, making use of branches up to 25 cm (80 ft) and fallen trunks to travel about the wood.

House mice are found in a wide range of habitats throughout the world from Antarctic islands to coral atolls and in Britain they range from cold stores to centrally heated buildings as well as offshore islands and arable land after harvest. They make runways similar to, but smaller than, those of **bank voles** and **wood mice**, which are easily seen as well defined routes through dusty places. Nests are made in houses behind skirting boards, under floors and even in ducting for cables and pipes. In well used situations, such as long pipes and beams, they leave 'pillars' (free-standing columns) made of urine and faeces as evidence of their frequent use. Dirty 'grease' smears are found on runways and semi-circular 'loop smears' of grease, smaller but similar to those left by rats, are commonly found, where obstructions interfere with easy travel, for example on the sides of beams, along roof joists. Nests are made, as might be expected, of any material which is to hand; they will shred paper and other material less finely than the **wood mouse**.

Away from man's influence **house mice** construct burrow systems similar to those of **wood mice** with a number of nest chambers and several exits to the outside.

Mouse boxes

Small mammals, especially the arboreal **wood mice** and **yellow-necked mice,** may make nests or simply rest in bird boxes or in special boxes originally designed as nesting or hibernation boxes for common dormice. These special boxes are similar to bird boxes but with a 26 mm hole in the back facing the tree trunk and kept away from the tree by an oversize base board and a 25 mm batten at the top. The mice climb the trunk and get in at the side of the box by climbing around the trunk from the side. Sites should be at least 1.5 m above ground and the more branches connecting at a low level with other trees and shrubs, the better.

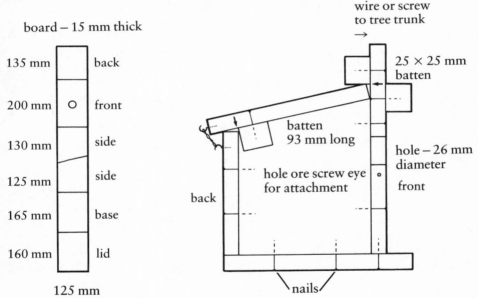

board – 15 mm thick

135 mm		back
200 mm	O	front
130 mm		side
125 mm		side
165 mm		base
160 mm		lid

125 mm

wire or screw to tree trunk
→

25 × 25 mm batten

batten 93 mm long

hole ore screw eye for attachment

back

hole – 26 mm diameter

front

nails

Mouse/common dormouse boxes are easily made at home and should be sited at least 1.5m above ground level.

Small mammal communities

Small mammals form characteristic groupings in different habitats according to the food and cover available; they are also influenced to some extent by the other species that are around. **House mice** cannot usually co-exist with **wood mice** unless they are living in human habitations; **bank voles** and **short-tailed voles** may exclude each other from their own preferred habitat in the breeding season and **wood mice** may avoid habitats with high densities of **short-tailed voles**.

It is assumed that the other small mammals, not covered in detail in this book (mole, pygmy shrew, common shrew and water shrew) will occur in most habitats. This broad generalization needs qualification as the mole, for instance, may be limited by a suitable depth of soil for burrowing and is uncommon, probably because of lack of prey, in coniferous woodlands, moorlands and sand dunes. Pygmy shrews and common shrews are found wherever there is enough ground cover; the latter is usually the more abundant of the two in grasslands and woodland but it is usually less common than the pygmy shrew in moorland. The water shrew, which is common near unpolluted watercourses, is found only sporadically in many other habitats.

Most work on small mammal communities has been done during the development of woodland from grassland, in clear-felled woodland, or coppiced habitat, as part of forestry management. This may seem rather artificial but it has revealed much about the small mammal communities generally associated with these habitats.

In one instance a large plantation which started as an old field, **short-tailed voles**, **bank voles** and **wood mice** were abundant after three years and **harvest mice** were found in small numbers as well. As the trees grew and shaded out the grass the **short-tailed voles** declined, leaving only **bank voles** and **wood mice** by the time the plantation was eight years old. Similar changes would be expected to occur in deciduous woodland over a similar or longer habitat succession and in southern England **yellow-necked mice** might be present as well, as the wood matures.

In woodlands that are planted for the first time in habitats with little grass cover or after clear-felling, the **short-tailed vole** population may be very small and never develop much before the canopy closes. The size and distribution of the **bank vole** population will depend very much on the type of ground cover available – bracken, bramble, tree brashings or heather suit them.

In a 'mixed coppice with standards' deciduous woodland studied in south-eastern England selected areas are 'coppiced' in a rotation of up to thirty years. This means that the standard timber trees such as oaks are left alone and the new growth is cut from the bases of the 'underwood' trees and shrubs. The newly coppiced (cut) young growth of hazel and many other species is used for hurdles, fencing, etc.

The coppice system, with various patches of the wood in different stages of coppice regeneration, provides an ideal situation to study the small mammal communities associated with each stage. In the year of coppicing **wood mice** are abundant and **harvest mice** colonize the rapidly growing tall-stemmed ground vegetation; **bank voles** are not as common as in mature woodland and **short-tailed voles** and **yellow-necked mice** are present but scarce (see table, p.64). By year 3 the **bank voles** have become dominant at the expense of the **wood mice** and the **harvest mice** and **short-tailed voles** have declined to low levels. By year 10 following coppicing the **wood mice** return to dominate the community; there is an increase to low levels of the **yellow-necked mice** and a reduction of the **bank vole** population together with the virtual disappearance of **short-tailed voles** and **harvest mice**. Later years show the maintenance of a 'mature woodland' type of community which is dominated by **wood mice** with

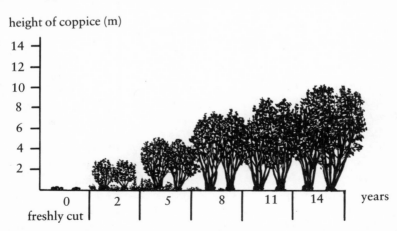

height of coppice (m)

In coppiced woodland the vegetation structure changes as the shrubs grow. Note the well developed low vegetation in the early years.

lower densities of **yellow-necked mice** and **bank voles** (and common shrews). Again the proportion of **bank voles** present is dependent on the ground cover available and the high level of **bank voles** in the 170-year-old woodland is simply due to the loss of trees and their shading canopy which allows glades to develop with more bracken and bramble for food and cover.

Relative abundance of small mammal species (% total numbers) in 5 ages of mixed coppice (mainly sallow, hazel, ash and birch) at Bradfield Woods, Suffolk, and in an oak woodland in southern England. (After Gurnell, Hicks & Whitbread, 1992).

| Species | Age of coppice (years) | | | | | |
	1	3	10	20	30	170yr oak
Short-tailed vole	3	2	<1	–	–	<1
Bank vole	10	37	19	16	10	41
Wood mouse	40	18	52	55	59	41
Yellow-necked mouse	2	2	10	11	14	4
Harvest mouse	20	5	<1	–	<1	<1
Common shrew	23	25	17	17	14	13
Pygmy shrew	<1	7	2	2	2	2
Water shrew	<1	2	<1			<1

Yellow-necked mice in woodland

Manchester University, under the direction of Dr Derek Yalden, has for a long time studied a woodland in Gloucestershire which has high proportions of **yellow-necked mice** *in the small mammal community. The woodland is thought to be especially favourable for this species because of its high productivity of seeds and fruits, especially beech and yew, which keeps both* **wood mice** *and particularly* **yellow-necked mice** *going through the winter.*

In the studies (see Figure), which started in 1968, the **yellow-necked mice** *were initially twice as numerous as* **wood mice**, *and* **bank voles** *formed 50% of the catch. However, from 1973 onwards the* **bank voles** *declined and from 1976 the* **yellow-necked mice** *declined as well, leaving* **wood mice** *to predominate in almost every year since. The* **wood mice** *have now become 2-3 times as numerous as* the **yellow-necked mice.**

The reasons for these changes in the small mammal community are not certain but very good arguments have been put forward. It is suggested that the loss of some deciduous woodland to coniferous plantations, the felling of a number of yew trees (highly nutritious regular producers of seeds) and the loss of the early seeding elms to Dutch elm disease have reduced the supplies of seed food for the **yellow-necked mice** which specialize in seed-eating. The situation was exacerbated after 1981 when the spaces opened up by the dead elms allowed wet snow to topple a further seven mature yew trees (yew seed is not poisonous to mice). The loss of elms may have helped the decline in the **bank vole** population in the late 1970s and 1980s by removing a green leaf food supply but more likely is the loss of ground cover as a result of shading as young trees, planted in the 1960s, matured.

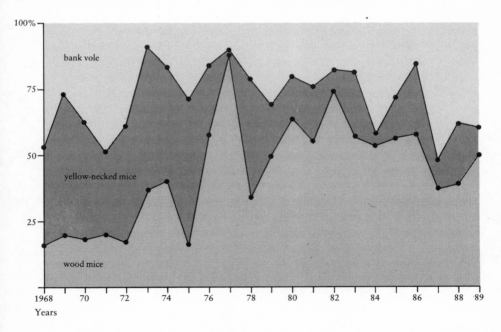

Percentage of each rodent species in a Gloucestershire woodland. Data from D. W. Yalden and R. F. Shore (1991), Journal of Zoology, London, 224, 329-32.

Food and feeding

Mice and voles show a wide range of feeding habits. **Wood mice, yellow-necked mice, house mice** and **bank voles** commonly eat 4-5 g of concentrated food each day, whereas **short-tailed voles** eat about 25 g (1 oz) of grass daily. The **wood mouse, yellow-necked mouse** and **house mouse** are generalist seed and insect eaters whereas the **harvest mouse** is a more selective seed eater. The **bank vole** is a generalist herb, leaf and seed eater and the **short-tailed vole** specializes in eating grass.

All the species are able to adapt their diets according to the availability of particular food items, which is partly why they are so successful. In times of insect abundance **wood mice** and **yellow-necked mice** will switch from their normal seed and fruit mixed diet to take more insects. Their bodies enable them to take more food when they need it: the alimentary canal adapts to the needs of the animal; the length of the gut and the development of the gut wall increase during pregnancy, and especially in lactation, which are both particularly energy-demanding periods of life. **Bank voles** will even eat their own faeces at any time, a common phenomen observed in plant-eating mammals like a rabbit, in order to increase the intake of protein and vitamins. These nutrients are released by bacterial digestion but only partly absorbed in the lower part of the alimentary canal. The appendix is a vestigial part of the gut in humans but in small mammals and many herbivores this becomes a large caecum where bacterial digestion takes place.

Wood mice and yellow-necked mice will eat invertebrate food as well as seeds and fruits.

Diets in different habitats

Wood mice in deciduous woodland usually eat seeds and fruits for most of the year but will eat more insects and leaves in the spring and summer. In coniferous woodland insects may make up to 50% or more of the diet, compensating for the lack of seeds and fruit, especially in the spring.

In arable land weed seeds, spilt or growing cereal seeds, earthworms, insect larvae and the remains of past crops make up much of the diet. In fields ploughed for winter wheat much sown and shed grain is taken during the autumn and winter with some insects, earthworms and weed seeds; in spring and early summer weed seeds and grass flowers and seeds predominate. In sugar-beet fields before autumn harvesting mainly weed seeds and some cereal seeds are taken; after harvesting and ploughing this changes to insect larvae and the remains of sugar-beet roots in late winter.

In gardens it has to be admitted that **wood mice** will take crocus flowers, bulbs, apples, peas, beans and even tomatoes!

The feeding habits of the **yellow-necked mouse** have not been studied in detail in Britain, but in general they appear to be very similar to those of **wood mice**. Studies in Europe show that they are more efficient than wood mice at extracting beech seeds but less efficient with grass seeds, indicating that the two species may not compete over food as much as might be suspected, and explaining why yellow-necked mice are not often found outside woodland. **Wood mice** will test beech nuts before eating them and discard those that have no kernel. Wood mice from sand dune habitats (no beech mast) cannot do this.

Bank voles that were studied in an oak-ash woodland in Britain ate mainly seeds and fruits in autumn and winter with a large proportion of herb leaves and even tree leaves in spring and summer. Studies from Europe suggest that eating large amounts of tree leaves is unusual and in coniferous woodland fungi seem to take the place of much of the seed and fruit food. **Bank voles** and **short-tailed voles** appear to need more water or succulent food in their diet than the mice do.

It is difficult to generalize about the feeding habits of **house mice** except that they appear to prefer cereal grains to most other foods. In free-living populations insects appear to be a main source of food and they will eat them when available in houses too. On Foula, one of the Shetland islands, it was found that **wood mice** tended to have more plant food than insect remains in their stomachs while the opposite was true of the **house mice**. However, in general their diets are considered to be very similar and possibly the cause of direct competition between the two species in the absence of man.

Harvest mice are often shown holding on to the stems of cereals such as wheat and they will take the ripe grains from the ears and from the ground, leaving characteristic sickle-shaped remains. However, they will

What won't a mouse eat?

House mice in particular are renowned for eating almost anything that is available. However, some 'foods' may actually be the incidental swallowings from chewing through a substance which may lead to a food supply. Examples of bizarre foods/chewings are: plastic electricity cable, plastic piping, lead piping, aluminium traps, match heads (they lit the match) and candles. In the 1950s straws were pulled out of cardboard milk tops and licked for any remaining milk.

	yellow-necked mouse	wood mouse	bank vole
seeds and fruit			
animal			
herbs			
fungi			
other			

Diagram showing the percentage of the diet (indicated by the size of the black bars) made up by each food type in woodland small rodents.

also eat a wide variety of grass seeds, the shoots and leaves of grasses and herbs, fungi, insects and other invertebrates. It has been calculated that as much food must be consumed in a day by an 8-gram **harvest mouse** as would be eaten by a 30-gram **wood mouse**; this is because, being smaller, the harvest mouse has proportionally greater surface area that loses heat.

Short-tailed voles are grass, leaf and stem eaters and show quite marked preferences for particular species. They will, however, eat bark and roots and this is one reason for their unpopularity in conifer plantations. The kinds of grass they particularly like are some of the 'soft grasses' such as *Agrostis* and *Festuca* species which may be shaded out by the more erect grasses in what apparently looks like a favourable habitat. The table overleaf shows the Latin and common names of grasses and other plants known to be eaten in Britain.

Grasses commonly eaten by *short-tailed voles*

Latin name	Common name
Agrostis canina	Velvet bent
Anthoxanthum odoratum	Sweet vernal-grass
Arrhenatherum elatius	False oat-grass
Festuca rubra	Red fescue
Poa trivialis	Rough meadow-grass
Poa pratensis	Smooth meadow-grass
Holcus lanatus	Yorkshire fog
Brachipodium pinnatum	Chalk false-brome
Brachipodium sylvaticum	Wood false-brome
Zerna erecta	Upright brome
Dactylis glomerata*	Cocksfoot*

*Unpalatable in some areas

Grasses were found to make up 67% and mosses 20% of the diet of **short-tailed voles** in an abandoned grazed field planted with conifers in south-west England; rushes, grass seeds, liverworts and herbs were also taken. Grasses took up more of the diet in autumn and early winter than in summer. The obvious piles of chewed-off grass stems and leaves found in runway systems in grassland are discussed in Habitats, burrows and nests.

Some species of grass, like cocksfoot, are unpalatable to **short-tailed voles** in some areas but are eaten elsewhere and so some geographical variation, perhaps in the defence mechanisms of the grass, may be involved. Tufted hair-grass (*Deschampsia caespitosa*) is probably unpalatable as well, but it is tough and is often used for nest-building.

Feeding signs

A number of food items are eaten in characteristic ways by small mammals. Cereal grains are often only partly eaten (kibbled) so that the chewed remains, known as 'kibblings' are left by **house mice** and **wood mice**; mention has already been made above to the sickle-shaped remains of cereal grains left by **harvest mice**.

The way seeds and fruits are eaten is particularly useful in identifying the species which ate them.

Hazel nuts (see diagrams) eaten by **wood mice** have a round hole gnawed in the side with a ring of tooth marks surrounding it. If the hazel

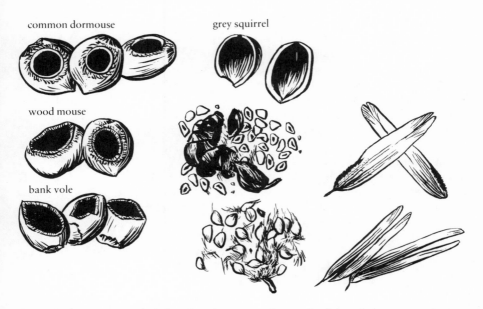

Characteristic marks left on hazel nuts, rose hips and ash keys by various rodents.

nut is opened by a **bank vole** the hole is clean-edged with no tooth marks around the outside. If the **common dormouse** has eaten the hazel nut the hole in the shell is definitely smooth and round inside and has oblique tooth marks around the cut edge. Red and grey squirrels split open the nut so that the shell is left in two halves with clean edges. This may seem fairly straightforward, but you will certainly have difficulty in telling signs of **yellow-necked mice** from those of **wood mice** – they have very similar feeding habits.

The seeds from ash keys are extracted differently by **wood mice** and **bank voles**. **Wood mice** chew a small hole in the side of the wing next to the edge of the seed and pull it out, whereas **bank voles** extract the seed by splitting the wing down the centre. However, this detective work must be done with care as birds will also extract the seed by splitting the wing down the middle; the only certain way to identify an ash key opened by a **bank vole** is to find it underground or in a feeding pile close to a tunnel opening, somewhere a bird is unlikely to have left it.

Wood mice tend to eat the seeds from some soft fruits whereas **bank voles** eat the flesh; thus when rosehip seeds are eaten by **wood mice**, they gnaw through to extract the kernel, and then leave the flesh uneaten; **bank**

voles on the other hand eat the flesh and leave the kernel. However, blackberry, elderberry and hawthorn berry flesh is readily eaten by both species. In addition, hoards of blackberry seeds may be found in birds' nests where probably a diligent **wood mouse** has sat chewing through the edge of every one to extract the kernel; hawthorn and elderberry kernels will also be eaten by **wood mice** but not by **bank voles**. **Bank voles** will strip the bark of trees like elder and eat the cambium below and they will also graze growing woodland seedlings.

Are mice and voles woodland vandals?

Bank voles and **wood mice** are well known for their selective grazing on tree seedlings and seeds. It has been calculated that these two species are responsible for the loss of 49-75% of ash fruit, depending on the size of the crop, by consuming the seed before it can germinate. In a study of beech woodlands in Denmark, small rodents were found to take only 1.7-4.3% of seed in heavy crop years, compared with 70-100% in other years. An experiment with lime fruits showed that 95% of the fruits (including the seed) were consumed, probably by **wood mice**, when sown at densities of 150/m^2 over areas of 25m^2 in woodland. **Wood mice** are so fond of acorns that they block up field drains by using them as stores in East Anglia.

Thus in some years small rodents have a marked effect on the survival of tree seeds. However, many seeds are 'scatter-hoarded', typically a few centimetres below the soil (they have been found in the walls of runways) and never found again. These seeds have a better chance of germinating than those left in the litter where they suffer from drought and frost effects. Against this is the fact that some of the caches will be found and the seeds eaten at some time in the future. Food stores and caches are common in the burrows of **wood mice**, **yellow-necked mice** and **bank voles**, particularly in the autumn and winter. Storing behaviour is brought on in laboratory **wood mice** when the daylength is reduced from 16 to 12 hours and with **bank voles** when the daylength is reduced from 12 to 8 hours.

The preferences of **bank voles** and **yellow-necked mice** for various seeds have been studied in Denmark. It was found from experiments with seeds from deciduous trees that the most popular were beech mast, acorns and hazel nuts, in that order; they would also take lime and elm. **Bank voles** also took some ash and alder while **yellow-necked mice** only took elm in addition (see table).

In woodland experiments where the same variety of seeds were left to be

Preferences in the laboratory of Danish small mammals for tree seeds

Species	Bank vole	Yellow-necked mouse
Beech	+ + + + + +	+ + +
Oak	+ +	+ + +
Hazel	+ +	+ + +
Lime	+ +	+ +
Ash	+	
Alder	+	
Birch		
Elm	+	+ +
Sycamore		

Number of + indicates relative preference

found under sheets of plywood, the three most-preferred seeds – beech, oak and hazel – disappeared most quickly and, apart from many hazel nuts, were eaten elsewhere. Little interest was shown in alder, elm or birch and the eventual losses of the other species were thought to be from digging rather than feeding. It is difficult to tell how these tests relate to a real feeding situation as the relative abundance of other food items may affect preferences for particular species. In ash woodland both **wood mice** and **bank voles** eat ash seeds throughout the winter if they are available, whereas in mixed deciduous woodlands, where acorns are often available, ash forms only a small part of their diets. Acorns form an important part of the diet of **wood mice, yellow-necked mice** and **bank voles** and, although poisonous to some larger animals, are not poisonous for mice.

Studies of seedlings and woody stemmed saplings (less than 20 cm/ 8 in high) planted in woodland and in the laboratory show that **bank voles** in particular damage or destroy lime saplings but do not touch birch saplings. Oak and beech seedlings are also damaged or destroyed but saplings, once the outer bark has begun to form, suffer little damage from voles. The saplings of elm (shoots), ash (bark) and sycamore (bark) suffered some damage but recovery was usually possible.

Lime saplings that were damaged often had their bark removed close to the ground and the lateral buds and several of the new shoots also eaten. Paired toothmarks could be seen along the edge of the remaining bark and double grooves seen across the cut surface of the remaining wood; measurements showed the individual teeth marks to be about 0.8-0.9 mm wide, as might be expected in **bank voles**. Comparison with damage to saplings in the laboratory showed that either the **bank vole** or the **short-tailed vole** could be responsible and that the **wood mouse** only occasionally removed buds and did not generally gnaw bark or bite through the woody stem. Damage in the field usually happened where **bank voles** were known to be, so the finger points to the bank vole as responsible for losses of saplings. However, the saplings of most of the species tested recovered after being damaged, producing leaves the next spring; the only species which did not recover was the lime, which was commonly eaten to below ground level. Again, the importance of **bank voles** in preventing regeneration of species like the lime will depend on the abundance of the saplings and other food supplies. It is estimated, however, that naturally germinating lime seeds rarely reach more than 1 to 5 per square metre in a wood and that quite low densities of **bank voles** could eliminate them.

Smells, squeaks and behaviour

Small mammals are often depicted as needing glasses, squeaking loudly and being smelly; true up to a point, but only partially true. Sight is not very important to mice and voles but we do know that **house mice** can orientate themselves towards visual landmarks and they can also discriminate between different shapes. Most species are nocturnal or live much of their lives in dim light conditions and so it is perhaps not surprising that sounds and smells are more important.

Mouse and vole squeaking is mostly silent to human ears as they can detect sounds at much higher frequencies (called ultrasounds) than humans and what audible squeaks (to humans) they do make are usually in fear when fighting or being handled. The **yellow-necked mouse**, for instance, is particularly vocal when handled and this is one of the many pointers used in differentiating it from the **wood mouse**.

Smells are one of the main sources of information about the environment for small mammals and, as with sounds, their smells are not necessarily very strong to the human nose. But even to us, **house mice** have a strong musty smell and **short-tailed voles** have a pungent cheesey-musky smell, produced by their urine. These smells serve to separate the species for the purposes of identification as **bank voles** smell very little in comparison and the **wood mouse, yellow-necked mouse** and **harvest mouse** hardly smell at all.

Calling mother
In **house mice** the mother is aggressive to strange mice near the nest after the young are born (a common characteristic in small mammals, to protect young from infanticide – see Numbers). The newborn **house mice** produce ultrasonic distress calls as do **wood mice, yellow-necked mice, harvest mice** and **bank voles**. When the temperature of the pups drops, as when the mother leaves the nest or when they crawl outside, those that have not yet developed their temperature control give ultrasonic calls so that the mother will return or retrieve them. Young **bank voles** and **harvest mice** make audible calls as well as ultrasonic ones. The reaction to these distress calls is so strong that a female **house mouse** will even move towards a loudspeaker producing calls from house mouse pups.

75

Ultrasonic calls are also made when young **house mice, wood mice** and **bank voles** are touched, even when they can regulate their own temperatures, and **house mice** will quieten down if presented with the smell of their own nest. Louder audible squeaks are only made by young **house mice** when they are handled more – an alarm call. Ultrasounds in response to cooling are made by **wood mice** until 19-24 days old and **bank voles** until about 12 days. Maternal responses must be common between species because it is possible to cross-foster them.

There are also stimuli from the mother which attract the young; juvenile **house mice** are attracted to the smell of lactating females, some of which comes from the mothers' faeces. This could serve to keep the young close to the nest as they develop to independence.

Adult communication

Ultrasounds are used by adult mice and voles during mating. In the **house mouse** the male makes ultrasounds which are particularly strong during his approach, nosing, ano-genital sniffing and initial mounting of the female. When a male has had some sexual experience he will make these sounds when in the presence of a female, her urine, faecal odours or vaginal secretions. The female **house mouse, wood mouse** and **bank vole** will make audible squeaks when she is rejecting the male's advances! In other situations audible squeaks are only made when an animal is under attack by another. **Harvest mice** also make audible calls during courtship and mating.

Wood mice make audible and ultrasonic calls during aggressive behaviour, especially the aggressor. **Yellow-necked mice** make calls when threatened, and threatened **bank voles** make audible squeaks; male **bank voles** are also thought to make ultrasonic calls during aggressive behaviour. **Wood mice** and **yellow-necked mice** also emit ultrasounds when they are exploring, especially when they are in groups; and in **wood mice** the calls are associated with the movement of the dominant animal.

A message in a smell

Most research on communication by smell in small mammals comes from **house mice** but a substantial amount is known about **bank voles** and **wood mice** as well. Small rodents have a wide range of accessory glands in their reproductive system and skin glands which are capable of producing odours, and well developed noses and senses to detect them. Smells may be used in long-term communication with many other individuals through scent marks left by secretions from skin glands, in the urine or faeces. It seems likely that at a minimum most small mammals will be able to assess species identity, sex and reproductive status from these secretions; they will also give information on social and sexual behaviour. A generalized mouse or vole's potential sources of odour are illustrated below.

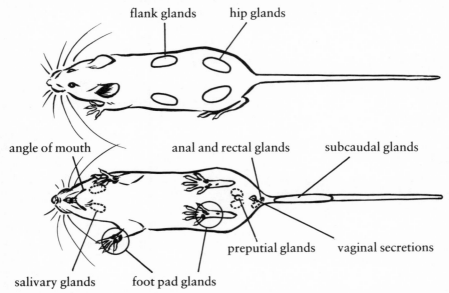

Generalized male/female small rodent showing some of the internal sources (dotted lines) and skin gland sources (solid lines) of possible odours used in communication. Note that urine is used as well.

Aggressive postures in the short-tailed vole; the dominant (on opposite page) *is threatening the subordinate* (above), *which is retaliating.*

In the **house mouse** odours have a strong effect on aggressive behaviour. Territorial males will attack intruding males but not adult females or juveniles. There appears to be an inhibitor of aggression present in juvenile and female urine. Males rubbed with juvenile female urine have much reduced aggression shown towards them by the territorial male.

We know that **house mice** can recognize individuals on the basis of their odour alone and can also recognize sex, and distinguish between different species; both males and females are attracted to the odours of the opposite sex. The sources may be from glands on the foot pads on the soles of their feet (!), accessory reproductive glands, preputial glands, salivary glands, urine and faeces, among others! They also have glands in the corner of the mouth, ears and anus. The salivary glands under the jaw (*submaxillary*) are larger in males than females and the saliva is used in social communication.

Male **house mice** have enlarged preputial glands with ducts leading to the tip of the prepuce and their secretions appear to be important in sexual and aggressive behaviour; females have equivalent, but smaller, clitoral glands, the functions of which are uncertain. The secretion from the preputial glands is attractive to females, especially if they are sexually experienced and not pregnant. Dominant **house mice** have larger preputial glands than subordinate ones and it is possible that the secretion is used in new environments to mark their territory or establish dominance.

The urine of **house mice** is very important in social communication and urine marking of the environment is well developed, as it probably is in most small mammals. Marking is done by placing spots of urine

particularly around the edges of their home range or territory or on prominent or novel objects; males will mark at a rate of 100-200 marks/hour whereas females will do it at 25-95 marks/hour (see Mouse territories). Males urine-mark more in the presence of females than with males and females mark more in the presence of males. Some females have been found to mark more frequently just before and during oestrus when they are receptive to mating. The prepuce of the male **house mouse** is adapted for marking as it has brush-like hairs on the tip and inside there is a space which acts as a reservoir for the urine.

Other odours in **house mice** include those from the vagina which are used by males to test if the female is in oestrus (ready for mating) or not.

In **wood mice** and **yellow-necked mice** there are glands which are potentially odour-producing in the corner of the mouth and lips as well as preputial (clitoral) glands, anal glands and a subcaudal gland on the lower surface of the first third of the base of the tail. The subcaudal gland is larger in the male than in the female and very large in male **yellow-necked mice**. It is curious that no marking behaviour is associated with these subcaudal glands but a milky secretion is easily expressed by gentle pressure and so it may be distributed during normal movements when the tail is in contact with the ground or other objects. Chemical studies of the secretions from the subcaudal gland suggest that odour from it may carry information on species, geographical population group and sex, and between adults and juveniles, but no behavioural tests have confirmed these proposals.

Wood mice appear not to deliberately mark their environment with

urine or faeces and no latrine piles of faeces are found; however, urine and faeces are deposited throughout the home range or territory and may still be used to signal occupancy to others. In **harvest mice** it is also thought that both faeces and urine are used for marking the home range.

Bank voles have anal glands, preputial glands, flank glands, and specialized glands in the angle of the mouth and lips, and in the pads on the soles of their feet. The preputial glands are larger in males than females, larger in adults than sub-adults and vary in size with season. Dominant males have larger preputial glands than subordinate ones. The flank glands are possibly associated with flank scratching with the hind foot (as with water voles) where secretion from the gland is deposited on the ground with the foot. Flank scratching (and defaecation, which is probably associated with the deposition of anal gland secretion) are performed most frequently by dominant animals and these behaviours are most commonly observed in aggressive situations. Females also use flank scratching and defaecation in aggressive situations; in addition to urine marking, they may be involved in territorial defence.

Bank voles mark new objects placed in the environment, such as traps, with urine: this is probably a major source of social information as it is also carried out when interacting with other bank voles. Sexually mature males leave urine trails, whereas females and immature individuals leave 'puddles'. Preputial gland secretion is commonly left with the urine, probably giving the streaks their brown appearance, and it is thought that sex attraction, indications of dominance status or territoriality may be communicated. Female **bank voles** are more attracted by the odours of dominant males and dominant males urine-mark more frequently than subordinate males. The marking urine contains a chemical called

 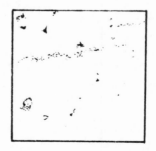

Urine trails of sexually mature male bank vole (left)
and female/immature bank vole (right).

hexadecyl acetate which is produced by the preputial glands, the production of which is dependent on male hormones. This chemical is apparently implicated in the assertion of male dominance as juvenile and subordinate adult males avoid traps smelling of hexadecyl acetate, whereas dominant males and mature non-subordinate males are attracted to it. Females were indifferent to the hexadecyl acetate odour and therefore the attraction of females to male marking urine is probably dependent on other chemicals.

In **short-tailed voles** there are anal glands, preputial (clitoral) glands, and hip glands in adult males which are much larger than those of females. The hip glands become enlarged in the breeding season. These voles are notorious for depositing large aggregations of faecal pellets at 'latrine sites' along their runways and it has been shown that anal gland secretions coat the faecal pellets as they are voided. The reactions of short-tailed voles to odours have been studied in relation to trapping and aggressive behaviour. They will enter traps tainted with their own odour or that of the opposite sex more often than other traps and once aggressive behaviour has taken place an individual will avoid the odour of the same species.

So small mammals use calls and smells to transmit social and sexual messages, and the small mammal's world is presumably dominated by these influences rather than by sight. **House mouse** eyes appear to be designed for low-light vision, and they are not very sharp – this probably applies to the other small rodents. Colour sensitivity is not thought to play a large part in mouse vision.

Observations of behaviour in small mammals mostly come from laboratory studies; it would be difficult to study many aspects of this subject in the wild. So such observations need to be judged carefully and if possible confirmed by observations in the wild. All the information about smells in relation to reproduction in the house mouse still remains to be confirmed by reports on wild animals (see Little breeders).

Numbers

Small mammals are renowned for reaching plague proportions. This is not very obvious in the more secretive woodland and grassland species, but still happens occasionally; it is so apparent in **house mice** that control measures become necessary (see Of mice and men). Numbers also 'crash' to very low levels at times and this can have serious consequences for small mammal predators (see Predators and predation).

Cyclical fluctuations

The regular occurrence of these peaks and troughs in numbers of small mammals has fascinated population ecologists for most of this century and studies are still going on to try and understand the intricacies of these fluctuations. The **short-tailed vole** and possibly the **bank vole** are the only British species that show these fluctuations in a cyclical fashion with periodic peaks at 3-5 year intervals and times of decline, relative scarcity and increase in the years in between. Cyclical fluctuations are typically found in voles and lemmings living in large expanses of tundra, prairie and moorland grasslands in North America and in northern Europe and Asia;

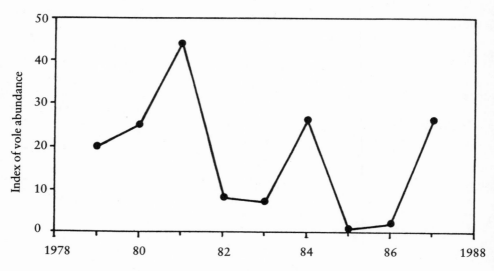

Yearly changes in the abundance of the short-tailed vole in south-west Scotland.

however, they are only well documented in northern and western Britain in **short-tailed voles**. Their cyclical fluctuations are particularly noticeable in young forestry plantations and upland grasslands. In south-west Scotland in 1981, for instance, there was a marked peak in **short-tailed vole** numbers and in 1985 there was a low: barn owl breeding (see figure and Predators and predation) was markedly affected.

In arable farmland in Cambridgeshire and Leicestershire a peak of **short-tailed voles** was recorded in the grassland along road verges and at field edges in 1981 and again in 1984.

Short-tailed voles are probably not as abundant as they were earlier this century when periodically numbers reached very high levels and they were notable forestry and agricultural pests. However, grasslands suitable for voles have diminished in area with the increase in the efficiency of agriculture, especially in the Midlands and south of England and this probably explains the lack of widespread vole cycles with 'plague' numbers in these areas (see Conservation).

Annual fluctuations

The woodland **wood mice** and **bank voles** usually show only annual fluctuations in density, the nature of which depends very much on tree masting (heavy seed crops). The usual pattern is for numbers to be low in spring, followed by autumn/winter peaks after breeding through the summer (see figure, p.84), but the increase in autumn and decrease in spring may be reversed in some years. The timing of the peak in numbers in the autumn may depend on the food available; in poor seed crop years in woodland (and with **wood mice** in habitats with poor food supplies, such as arable land) the peaks are relatively early. However, after a good autumn seed crop the numbers of **wood mice**, **bank voles** or **yellow-necked mice** in woodland may continue at high levels into the spring and summer and then decline to the next winter and spring, probably because breeding is reduced under high-density conditions (see pp.45, 88). In the New Forest **wood mouse** and **bank vole** numbers were very low in 1982 after a cold winter and a failed acorn crop and this pattern seemed to be common for these small rodents throughout the country. However, cold weather does not necessarily lead to poor survival over the winter for these species. In the cold winter of 1962-3 a good acorn crop and winter breeding were observed at Wytham Woods, near Oxford, and summer numbers of **bank voles** and **wood mice** were not greatly reduced.

The importance of winter food supplies to the survival of animals to the following summer has been borne out in an experiment where **bank voles**

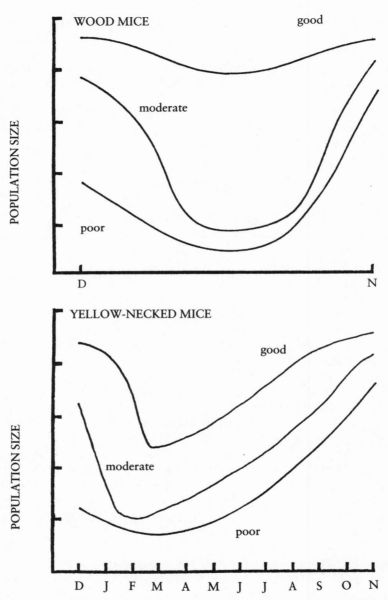

Idealized patterns of annual changes in numbers of wood mice and yellow-necked mice in woodland in years following poor, moderate and good seed crops.

living in an ash woodland were given extra ash fruit one winter. These **bank voles** survived to the spring and summer much better than those in a nearby population without the food. Similar results have been obtained by giving supplementary cereal food to **wood mice** in mixed deciduous woodland. For the **bank voles** the influence of the good winter food supplies often lasts until the following winter: numbers increase to higher levels due to the large number of animals surviving the first winter and breeding.

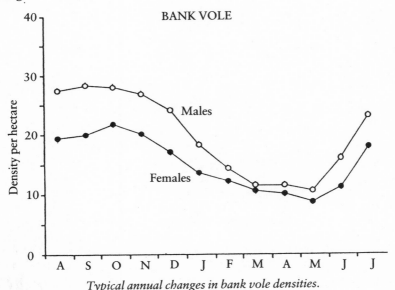

Typical annual changes in bank vole densities.

Yellow-necked mice have similar population fluctuations to **wood mice** but they reach a peak at an earlier time in the winter than **wood mice** and then decline earlier, increasing earlier to the autumn as well.

Harvest mice in grassland show very marked annual fluctuations, and may reach extremely high densities in autumn and winter from apparently low summer levels, see p.86. However, they move in the upper layers of the tall vegetation/grassland habitat for much of the summer and so the live traps placed on the ground in surveys may fail to record them.

House mice living in natural habitats are also influenced greatly by food supplies and by cold temperatures. The increases in **house mice** in beech woodlands in heavy beechmast years in New Zealand have already been described (see Little breeders). Population studies of the **house mice** living

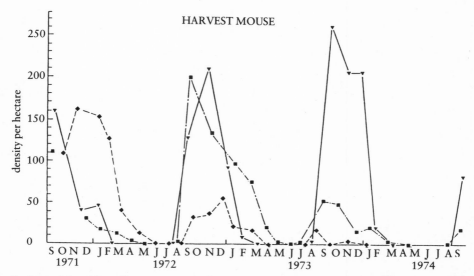

Estimated densities of harvest mice in three areas of rough grassland. Note that numbers in summer may be underestimated because they are not easily trapped at this time.

on Skokholm island off South-West Wales show the strong influence of winter temperature on numbers; it was found that the numbers present at the end of the breeding season (April-September) in the autumn were strongly related to the average temperature in the previous February (taken as an index of the coldness of the winter weather).

How long does a small mammal live?

An accurate answer to this question for wild mice and voles is very difficult to give but there are indications from limited data obtained from the wild populations. It is unusual for any small mammal living in the wild to survive for more than one winter; normally an individual is born in one season, may breed in that year or simply stay sexually immature over the winter and mature to breed in the following spring and summer; the vast majority of animals that survive one winter will have died before their second winter. However, many small mammals die very young, soon after they emerge from the nest, and it is possible that many more die before weaning.

The average life expectancy of a **house mouse** at birth in the wild is about 14 weeks; in **wood mice** from around 4 weeks of age (soon after weaning and independence) the average length of life is about 8-15 weeks.

A fast life

Mice and voles have a high surface-area to volume ratio in comparison with most mammals and thus lose heat very quickly as well as warming up very quickly. The smallest mouse in Britain, the **harvest mouse**, is very susceptible to low temperatures, and adverse weather conditions at the end of the breeding season cause the deaths of many juveniles; sudden drops in temperature and hard frosts also cause deaths in adults.

The fur of mice and voles is very important in insulating the body and it accounts for 30-40% of the total insulation of the **house mouse**. House mice in cold stores have fur which is 20-80% better! The body temperature of **house mice** and probably of most of the other species of mice and voles is in the range of 34-40°C whatever the external temperature (but there are exceptions). Once the temperature drops to 30-32°C or below the mouse starts to warm itself up by huddling with others in the nest or by becoming more active. If neither of these works then the blood supply to the body tissues near the surface is restricted so that heat is conserved and a number of physiological mechanisms come into play to generate heat. The mechanisms are complicated and include shivering from muscles and non-shivering production of heat from specialized areas of the body; these areas are mainly the brown fat tissues which are between and under the shoulder blades and over the breast bone and also the heart and skeletal muscles. **House mice** from S. Georgia in the Antarctic have over twice as much brown fat between the shoulders as those from Hawaii to help them cope with the lower temperatures. The other way in which mice cope with cold is to go into torpor. This is simply a state of suspended animation at a lower temperature (hypothermia) and is brought about by food deprivation and cold in **house mice, yellow-necked mice** and **wood mice**; the **wood mice** appeared to be about 10°C lower in their body temperature, huddled in a nest, and could return to normal quite quickly.

House mice have heart beat rates of 300 beats/minute in juveniles and 600 beats per minute in adults, 8½ times the human adult heart-beat rate of 70 beats per minute. These rates may be increased during excitement but they are still low in comparison with the heart beat of the common shrew, which at times of alarm, excitement or great activity changes from 88 beats a minute to 1,320 beats/minute!

What happens to young in the nest?

A study of **bank voles** born in underground nest boxes in Poland showed that 31-53% died before reaching trappable age, soon after weaning, and this may be quite common with the other species. Infanticide is common in rodent species and **wood mice** have been shown to do it in the laboratory and probably in the wild as well; in enclosure studies female **wood mice** were only able to rear litters successfully if they prevented other individuals from gaining access to the nest. In the laboratory it was the individuals that were not breeding (males and females that had not mated recently) that were the most likely to kill litters in the nest. For the male attacker the advantage is that the female may become more quickly receptive to mating (as is seen in other species); also he may lessen competition between his possible progeny and those killed. In addition, when numbers are high and many males and females are not breeding, the rate of infanticide may go up, thus helping to limit the population increase (see below).

Mouse family planning

The numbers of rodents are limited by something called 'reproductive inhibition': females do not breed because of chemical and behavioural influences when living at high density. However, the fact that plagues, particularly of **house mice**, occur shows that in this species at least the mechanism of reproductive inhibition only works at very high densities, especially if the feeding and nesting conditions are favourable. Indeed, it seems that good food supplies during the winter can override these influences in **wood mice, bank voles** and probably **yellow-necked mice** as well, for as long as the good conditions last.

With **bank voles** that live in confined situations such as islands it has been shown that adult breeding females occupy territories (defended areas from which other females are excluded); these areas are mainly occupied by females that have overwintered and females born early in the year. If some of these early-born females were removed more young born later in the year were able to take territories, mature and reproduce than would have done otherwise. This experiment suggests that the number of breeding females is determined on the island by the number of territories available and so population increase of the breeding females is strictly limited. Because of the limits on movement due to its being an island, there can be no population expansion, and young females remain as non-breeders; in continuous habitats many individuals would disperse to breed elsewhere.

Territoriality in breeding females seems to occur in **wood mice** as well (see Mouse territories). Dr Ian Montgomery in Northern Ireland has found that when the population of wood mice is high, again, there is a decrease in reproduction. It is suggested that female numbers are limited by finding territories in which to breed and that males are limited by food supplies and, more importantly, by finding a female to mate. At high densities fewer females are able to breed, fewer males and females join the population because of lower food availability and low breeding opportunities, and fewer young survive because the adults probably isolate them and prevent them from feeding, and there is also an increased possibility of infanticide (see above).

In **short-tailed voles** the opportunity for regulating population numbers seems to be less. The males are territorial and a number of females, more as the season progresses, have home range areas within the males' territories and their movements are not restricted. The males probably only form a social hierarchy under high density conditions and so other males can move around freely. The animals can move away easily and the populations appear not to inhibit reproduction as much as the **wood mouse** or **bank vole**, allowing numbers to reach very high levels.

In **house mice** the populations may be colonial (with many individuals sharing the habitat) or territorial (living in exclusive areas apart from individuals or groups of the same sex – see Mouse territories) in any type of environment. A variety of mechanisms exist that could potentially limit numbers. There are three main types of social organization, but they do not necessarily depend on population density. In enclosed situations hierarchies based on dominance are usually found; one male is dominant over several males and females and he will patrol and defend a territory and father most of the young. In group (shared) territories, often found in relatively stable populations in man-made habitats, several males share a territory on an equal basis. In an individual territory situation the resident mouse excludes other mice of the same sex from its home range. In low density situations, individual or group territories are likely to be found whereas in high density situations a hierarchy is more likely.

In studies of mice living in enclosures (at relatively high densities) **house mice** limit numbers by a variety of mechanisms. Breeding may be inhibited by loss of fertility, 'reproductive inhibition' (as we saw on p.88) may occur in females, or infant survival may decrease because of infanticide, poor nest structure or because the females are stimulated to retrieve other females' young too often rather than feeding their own. **House mice** adjust their numbers according to the prevailing environmental conditions.

Predators and predation

Life is often short for small mammals such as mice and voles: they are the staple diet of many predators both on the ground and in the air. If their numbers rise, so do those of the predators, who breed better and achieve better distribution. Equally, the more predators there are, the fewer and scarcer are their prey, especially if alternative prey are not available.

It goes without saying that before you can eat your mouse, you must catch it. **Short-tailed voles** live mostly under grass cover and so will not be vulnerable to birds until they venture into shorter grass, which perhaps happens if they wish to feed or explore. The predators that rely on seeing or hearing their prey become accustomed to feeding when the prey are active. It has been found that kestrels hunt in bouts coincident with the active phase of the **short-tailed vole's** daily rhythm (see Mouse territorities) and tawny owls only hunt at night in woodland when **wood mice, bank voles** and **short-tailed voles** are all active. In contrast the weasel, especially the female, can enter most small mammal burrows and can therefore hunt whenever it is hungry, detecting the prey by ear and eye and to a lesser extent by smell.

The species taken may vary according to the density of the prey and alternative prey, according to habitat type, weather and the behaviour of the predator. If a particular species is relatively scarce, a predator may 'give up' looking for it and concentrate on another species which is easier to find because it is more abundant; when mice and voles are at a low density tawny owls concentrate on worms and even frogs. Habitats affording cover from aerial predation protect the inhabitants. An example possibly illustrating this is the feeding of barn owls on Skomer island, off South-West Wales. The owls feed on the **Skomer bank vole**, the **wood mouse** and the common shrew; studies of the ages of the vole prey show that mainly young individuals were taken during one summer season. This may merely reflect the number of young around it or may reflect the fact that the younger animals have to take up residence in the more marginal habitats with less cover, and are therefore more easily caught.

Rain is rather good news for mice. In wet weather **wood mice** make much less noise when moving over the woodland floor leaf litter. Owls take fewer of them when it's wet than they do under dry conditions

Tawny owls perch on branches and swoop down on their prey.

because they need to hear their prey as well as to see them to be able to home in on them accurately. Even if the owls can hear the mice, the rain drops dripping on the leaves distract them as well. On dark nights owls have more attempts at catching prey but their efficiency at catching is worse than on bright moonlit nights; on the other hand moonlit nights have fewer animals moving about but enough light to see them by.

Generally speaking there are two main types of predator: the *specialists* are well adapted to prey on small mammals, and their breeding, density and distribution are greatly affected by the abundance of their prey. Specialist predators such as the weasel and tawny owl remain in one area and about half of their diet is made up of small mammals; both are influenced by declines and gluts in the numbers of small mammals available. Some specialists must move around to take advantage of areas of high prey density – such as some birds of prey in the non-breeding season. *Generalists* such as the cat, fox and stoat take a wide variety of prey (see p.19) and though they may have to change their choice of food, their breeding and numbers will not be greatly affected by declines in numbers.

Predators may select particular types of prey because they encounter them more often, because they identify them as prey more often, they are able to kill that type more often, or because they choose to eat that type more often (it may taste better to them). A study of tawny owls showed that in deciduous woodland they appeared to catch **bank voles** in the same proportions as they were present in the different cover types in the wood (as was established by live-trapping); however, they caught **wood mice** proportionately more often than the traps did in open ground and less often than the traps in areas of dense cover. In the open ground the owls took proportionately more **wood mice** than **bank voles**, presumably because mice use the open habitats more. Overall, they took relatively more **wood mice** than **bank voles** in comparison with the numbers available from trap records, again presumably because the mice use open ground habitats more often than the voles, and because it is easier to catch a mouse in the open than a vole in thick cover; but the sexes and size classes of both species were in the same proportions as shown in the live trapping. (Note: owl pellets are illustrated on p.29).

Observations of fox feeding habits indicate that they prefer particular prey over others. **Short-tailed voles** are generally eaten or cached (buried for later use) and if found again commonly eaten. **Wood mice** and **bank voles** are rarely eaten, although they may be cached, and, even if shrews are caught, they are rarely eaten and commonly discarded without being cached. Feeding studies indicate that foxes only eat shrews when cubs are present. In contrast, weasels will eat any of the mice and voles above in the order they were caught and only reject shrews, presumably because they are unpalatable. Common shrews make up only 5% of the tawny owl diet, but whether the owls find them unpalatable is uncertain.

Studies of tawny owls, barn owls and many other avian predators show that a shortage of small mammal prey has a dramatic effect on their breeding. In south-west Scotland it was estimated that there were about 140 breeding pairs of barn owls in a poor vole year (1985) and about 320 when **short-tailed voles** were abundant (1981). Despite nest sites being available in the poor years, some owls failed to breed. Exactly the same thing happened with tawny owls in woodland; in a study of the owls in Wytham Wood, near Oxford, when **wood mice** and **bank voles** reached a low density in 1955 (a year when the loss of rabbits, following the epidemic of myxomatosis, forced many other predators to take mice and voles and so compete for food with the owls) very few young were fledged and in 1958, when mouse and vole numbers were even lower (for uncertain reasons), none of the tawny owl pairs laid eggs at all. It seems that in times of low rodent density the breeding pair of owls is too undernourished to breed. This may not be too bad for the owls as they live for a long time (4-9 years on average). There are relatively few vacant territories to fill; so many fledged young have to emigrate in order to find a breeding territory, and many die trying to do this.

The population numbers and breeding of many other birds of prey seem to be affected by vole abundance: the kestrel, long-eared owl, short-eared owl and hen harrier all are strongly influenced. The diets of the four larger owls are shown below. It can be seen that there is some separation by diet between the species; the tawny owl lives mainly in deciduous woodland and it takes more earthworms and beetles while **short-tailed voles** make up a relatively small proportion of the diet compared with the other owls. The long-eared owl is more of a coniferous woodland/farmland/moorland dweller and **short-tailed voles** form an important part of its diet; the birds taken by the long-eared owl tend to be larger species than those eaten by the tawny owl. Short-eared owls are rodent specialists and are separated from the barn owl by habitat more than by prey, being more nomadic and having a preference for isolated tracts of country such as moorland, heathland, bogs, marshes and upland pastures. The barn owl is mainly a lowland species living in habitats such as mixed farmland, pasture bordering on moorland, young forestry plantations and rough grassland; it specializes in **short-tailed voles** and other rodents but, more than the other owls, it will take shrews when vole numbers are low.

Diet of four species of owl in Great Britain
Percentage by weight of each species in the diet

Prey species	Tawny owl	Long-eared owl	Short-eared owl	Barn owl
Common shrew	2.1	1.0	0.8	12.7
Pygmy shrew	0.1	0.2	0.3	1.2
Wood/Yellow-necked mice	12.5	14.9	7.0	10.4
Short-tailed vole	18.7	45.8	55.4	59.7
Bank vole	10.9	8.1	0.9	3.9
Brown rat	2.3	8.3	21.8	6.4
Birds	16.0	17.2	10.2	2.9
Worms	18.1	–	–	–
Beetles	0.4	–	–	–
Other	18.9	4.4	3.6	2.3

From Yalden (1985) Bird Study, 32, 122-31.

Weasels in woodland usually occur at low densities and fluctuations in small rodent prey are compensated for by changing to alternative prey such as birds. It has been shown that in years of low rodent numbers weasel predation on nest boxes containing nestling tits is much greater than in other years. This does not mean that the tits were the main alternative prey for weasels when small rodent numbers were low but just that they were affected more by weasel predation when rodent numbers are low. More data are needed on weasel diet in high and low rodent years to test this point.

Weasels living in Sussex farmland are greatly influenced by the density of **short-tailed voles**. They feed on many other small mammals including **bank voles, wood mice** and **house mice**, but not in large quantities, as well as small rabbits and passerine birds. **Short-tailed voles** feature more in the diet as they become more numerous (16% of a weasel's diet in a year when the field vole population is low, but 54% in a high vole density year). However, it seems that if weasels can't get enough of their favourite food (**short-tailed voles**), they turn to birds as substitutes on the menu, in preference to rabbits or other rodents.

The short-tailed vole has another important effect on weasels – the weasels have more young when there has been a glut of them. In the year following high vole numbers, weasel numbers increased; in a year of low vole numbers female weasels failed to breed. Thus the weasels were

High prey numbers lead to low prenatal mortality and high post-natal survival in stoats.

greatly affected by the vole population dynamics and, curiously, may have had some effect on the voles in return (see later). Similar effects of vole peaks on numbers and breeding have been shown in France and elsewhere; weasel numbers increase because adults that have survived the winter breed in both May and August/September and because the young of the first litter breed in the year of their birth. It has been calculated that a single female who has survived one winter under ideal conditions could produce 30 descendants by the end of the year; she would produce 2 litters of 6 young herself and if half the first litter was female these 3 could each produce a further 6 young.

Stoats have been introduced to New Zealand by man and in beech forests they include **house mice** (also introduced) in their diet. The beech trees have a heavy seed fall every 3-4 years and the mice and stoats show a strong reaction to this by increasing their numbers. The house mice breed more successfully than normal, breed longer into the winter and survive better, so that by the summer following a heavy mast crop there is a massive increase in the number of house mice present in the forests – much higher than in normal years. The stoats take advantage of this increase in food supply by eating the house mice and so their numbers increase, but by a different mechanism from that of the weasel described above. The stoat is unusual in its reproduction in that the females become pregnant soon after weaning and give birth in the spring of the following year. The embryos remain dormant for much of the pregnancy and the early embryos (blastocysts) do not implant in the uterus for 9-10 months with normal growth and development taking only 3-4 weeks before birth. The stoats' maximum number of young to be born is determined by the number of eggs fertilized at mating but in normal years many losses occur from the failure of blastocysts to implant and the resorption of embryos during pregnancy, as well as poor survival of the young. When house mice are plentiful none of these losses occur and many young stoats survive to independence in the summer following the heavy beech seed crop and the associated increase in house mice. Another knock-on effect of the good food supplies for the stoats is that the females produce more eggs and therefore have the capacity to produce more young in the following year; but the number actually produced, as stated above, depends upon the food supply during pregnancy over the following year.

All these examples show that predators ensure that they take full advantage of times of small mammal abundance by producing the maximum number of independent young by a wide variety of means. However, this increase in the production of young may be checked by other constraints, such as the type of social structure and the availability of suitable habitats to occupy before the breeding population is increased.

The hunter and the hunted

It is difficult to measure the effects of predators on prey but where studies have been done, some interesting conclusions have been drawn. In one

Short-tailed vole availability and weasel numbers

*After myxomatosis arrived in Britain in the early 1950s the rabbit population crashed and many grassland habitats that had been heavily grazed became suitable for **short-tailed voles**. This led to an increase in weasel numbers which lasted at least until the mid-1970s in some areas of Britain. The information on weasel numbers comes from gamekeepers' records and should be treated with some caution, but the association of the two events seems to be correct as peaks of weasel numbers were found in the records every 3-4 years, as might be expected with short-tailed vole predators.*

Weasel and prey.

Scottish plantation it was calculated that in late winter predation was the most important single cause of losses to a **short-tailed vole** population, but when breeding started in late spring the predators (weasels, stoats and short-eared owls) could not keep up with the increase in prey numbers. In the study of weasel numbers, weasel diet and **short-tailed voles** in Sussex farmland mentioned above (p.94) the weasels took more voles, after a slight delay, as the vole numbers went up, but the actual number killed could not be measured. Thus it is possible that the weasels were having an important influence on the vole numbers but the details of the interaction are uncertain.

In woodlands in Britain both weasels and tawny owls take large numbers of **wood mice** and **bank voles** each year. However, the influence of this predation on the small mammal populations has been very difficult to quantify. In Wytham Woods, near Oxford, information on tawny owls and small mammals was gathered by the late Dr H.N. (Mick) Southern and his co-workers from 1947 to the late 1960s and work on weasels was also carried out by Dr C.M. King. The work on the tawny owls estimated that they took 20-30% of the **bank voles** present in each two-month period and probably more of the **wood mice**. The weasels appeared to be less important predators as they were estimated to take an average of 8-10% of the combined small mammal population in the wood each month. However, it was concluded that at low rodent densities the combined effects of these two predators must be substantial as they would take a large proportion of the rodent population (although it was noted that weasels will move away to find alternative prey at very low rodent densities). At high rodent densities the predators took only a small proportion of the mice and voles so their populations were able to increase. Thus the general effect of the owl and weasel predation would be to exert more influence on rodent numbers as rodents decreased and to exert less influence as rodent numbers increased: something that is often found to be the case in relationships between predator and prey.

It is more complicated to sort out the effects of generalist and specialist predators on small mammals; one study compared the situation in a patchwork of forest and agricultural habitats in northern Europe with that in an area further south (in Sweden). In northern areas the predators are mainly avian and mainly specialists but in the south there is abundant alternative prey and the predators are a mixture of mammalian and avian specialists and generalists. In southern Sweden the specialist predators are the long-eared owl, kestrel and stoat, and the generalists are the fox, domestic cat, badger, polecat, buzzard and tawny owl. The conclusions of the study were based on a large number of assumptions but it was estimated that close to the total annual production of both the **wood mice** and the **short-tailed voles** was taken by the combined efforts of the predators, although some young must have been left to replenish their numbers. In this case, then, the predators had a very strong effect on the populations of both small mammals, preventing the **wood mice** and **short-tailed voles** from reaching very high densities or even showing a typical 3-4 year cycle of abundance in the case of the voles. When small mammals were at a low density the generalist buzzard, fox and cat could switch to the many alternative prey present (including water voles, rats and rabbits) and so remain to take advantage of the small rodents as soon as they increased in numbers again. In contrast, in the north of Sweden in continuous forest where the smaller number of predators were all specialists and fewer alternative prey were available, it was suggested that the numbers of predators followed the numbers of their small rodent prey and so 3-4 year rodent cycles were not prevented.

In Poland one of the most abundant small mammal prey species is the common vole (*Microtus arvalis*), a close relative of the **short-tailed vole** and it seems that intensity of predation on forest rodents is determined by the abundance of the common vole. When the predation (by tawny owls, foxes, martens and many other predators) of small mammals in agricultural and woodland habitats was studied it was found that in a year when these voles were abundant, only 33% of the woodland rodents (**bank voles, yellow-necked mice**, and other close relatives of the **wood mouse**) were taken and only 20% of the common voles. In contrast, when the voles declined, the predators took 95-97% of the woodland rodents and 89-100% of the common voles (note the similarity to the change by barn owls to shrews when short-tailed voles are scarce – see p.93). When the voles were abundant the combined predator force was very large and as the common voles declined the foxes, buzzards and cats joined the tawny owls and martens in taking the woodland rodents. It was therefore not surprising that the effect on the woodland rodent population was so intense!

Thus predation can exercise a strong control over numbers of small rodents, but in some cases it seems to be possible for the mice and voles to break out of the influence of the predators and reach very high densities.

Parasites and disease

Small mammals are, like all mammals, subject to parasitic infections and diseases, some of which may be passed on to man or to his domestic animals. This is another aspect of the economic damage that mice and voles can do and control to prevent public health problems is often necessary, especially with **house mice**. Parasites or disease may be important in reducing the large numbers of small mammals, but there is some evidence that they may even control numbers at very low levels.

Mice and voles are subject to fungal, bacterial, viral and protozoan diseases as well as helminth (parasitic worm) infections and flea, louse, mite and tick infestations. Those important to man are mainly diseases of the **house mouse** such as the bacterium *Salmonella* (food poisoning) which, surprisingly, is not very common as a result of contamination by mice in the UK but may be more common elsewhere. Other diseases possibly passed on to man are rickettsial pox (transmitted by a mite), rat-bite fever, leptospirosis (Weil's disease), and murine and scrub typhus, although most of these will only be spread in other countries.

Diseases that may be passed on to domestic animals are leptospirosis (see box), salmonellosis, helminth (worm) infections and fungal infections of the skin such as ringworm.

Obvious symptoms of disease are rarely seen in small rodents, although after outbreaks of 'plague' proportions of **house mice** or **short-tailed voles** animals are seen dead and dying, quite often from disease (mostly presumed but proved in some cases). In two enclosed populations of laboratory house mice which were prevented from increasing beyond 300 individuals by the removal of young, an infection of nematode (thread-worm) parasites was introduced and the effects observed. The parasite infection killed over 90% of the mice within 6 months in the two experiments and numbers did not increase to over 50 until an anthelminthic ('worm' treatment) was given. This indicates that parasites may be important causes of death and experiments with **wood mice** in 9 × 12 m enclosures suggest that high parasite loads are associated with mortality. However, in studies of free-living **wood mouse** populations, resistance to the parasites and the assessment of the causes of mortality make further confirmation difficult.

Parasites to look out for when handling mice and voles are the fleas which are relatively specific to the host but will jump onto alternative

Two bugs to avoid

Leptospirosis: a spirochaete bacterial disease (Weil's disease) caused by the strain which is usually present in the common rat. It is possible to catch this or milder strains of the disease from small mammals. Infection is usually through contact with urine-contaminated water or urine itself and it seems that transmission is through abraded skin directly into the blood stream (watch out if you are bitten and keep the wound clean – preferably wear rubber gloves when handling small mammals). Symptoms include fever, headache and possibly jaundice but they are usually from infection by rats and can be treated. A survey of people who had handled small mammals showed that none had been infected with the disease.

Lyme disease: a bacterial infection transmitted through the bite of a tick. Ticks will feed on most mammals and any tick bite should be looked on as a possible source of Lyme disease, although many bites will be harmless. Look for a circle of red inflammation of the skin around the bite which increases daily. The other symptoms are variable but may include headache, a stiff neck and 'flu' symptoms, and later on arthritis, meningitis, paralysis and even death! Treatment with antibiotics as soon as possible is the best remedy and any doubts about health after a bite by a tick (which stays attached to the skin for a long time) should be put to rest by visiting the doctor.

mite
(1 mm)

larval tick
(3 mm)

tick
(up to 13 mm)

louse
(2 mm)

flea
(3 mm or more)

hosts including man at times. A common home for a flea is the nest of the mouse or vole where the larva often develops into the adult; once the adult is warm enough it will jump onto a passing host and take a ride and a blood meal which is probably necessary before mating can occur and eggs released back into the mouse or vole habitat. The largest flea to be occasionally found on small mammals is the so-called 'mole flea' which is about 5 mm (¼ in) long and which appears to thrive on a wide range of species. For a short-tailed vole, this would be like a human having a mouse running over its body. Tick infestations are usually the larvae (six legs) or adults (eight legs) which engorge and drop off, usually being found behind the ears. Mites and lice move on and between the hairs eating skin or sucking blood and are very small but nevertheless very common and orange 'harvest mites' sometimes form clusters in the perineal region of mice and voles. All these 'travellers' appear to cause their hosts very little distress.

The mole flea is a common traveller with many small mammals.

Of mice and men

House mice are familiar pests in houses, usually got rid of by traps or poison bait; but the damage caused to domestic food is on a different scale from the damage they may do to industry and agriculture and in communal buildings; here great efforts may be needed to prevent further problems.

Infestations of cereal crops by **house mice** are not common in Britain, but in other countries, such as Iraq, up to 33% of the losses of corn crops is caused by them. It is likely that wherever house mice occur near cereal crops their food will include much grain and the damage is likely to be out of proportion to their need for food. This is because **house mice** only need 3-4 grams of food daily but they waste much more as they take the grain to pieces but only eat a portion. In addition, in situations where food for humans or livestock is stored much of it is fouled and made inedible unless it is cleaned. In the USA about 70% of the corn samples tested were contaminated with mouse droppings and even in 'cleaned' samples droppings still get through. This sort of contamination is not confined to foreign countries: in samples of grain from Scotland reported in the late 1950s there were 96 mouse droppings and 14 rat droppings in every 10 lb (4.5 kg) with 98% of the samples contaminated to some degree. Urine contamination presents a similar problem, but confirming it is much more difficult. Contamination of food stuffs by whole mice also occurs; a bizarre example of this was a perfectly sectioned animal once found in a loaf of sliced bread!

Damage by mice to industrial and domestic premises occurs in various ways, most usually by gnawing. The teeth of **house mice** (see Town mouse, country mouse) are very strong and they can chew through plastics such as those found in pipes and electrical cables (the latter with sometimes fatal results for the mice), lead and aluminium. They can also help to start fires by short-circuiting electrical equipment directly by making the connection themselves.

In a Birmingham office the computers were plunged into chaos when some local mice, which were resistant to poisons (see below) took a penchant to chewing cables!

There are problems in controlling mice that are causing damage: in domestic properties that have a large number of occupants there are frequently many sources of food, such as kitchen areas and communal dustbins, and few people are usually willing to take the responsibility for doing the nasty job. Modern buildings often have ducts which carry pipes and cables leading from one property to another and so **house mice** can travel around easily. Even old properties often have ready-made runways under floorboards or skirting boards so that access from one house to another is again easy.

What's your poison?
Old-fashioned remedies for eradicating **house mouse** infestations included the acute poisons yellow phosphorus and red squill (an extract of a powder obtained from a species of Mediterranean lily), which were banned in Britain in 1963. However, other acute poisons like zinc phosphide (added water or dilute acid causes the fatal phosphine gas to be emitted) may still be added to baits and used under certain conditions. Poisons such as these are not very effective because house mice, like rats, show 'bait shyness' – the animal feels ill and does not eat any more of the poison even if it is still attracted to the 'food'. To combat this the poison may be used as part of a contact dust which is placed where the mice will brush past it and lick it off their fur.

In the early 1950s anticoagulant rodenticides were introduced and were an immediate success. These poisons act by stopping the normal coagulation of the blood at minor cuts and many small internal haemorrhages from almost any part of the body so that the animal slowly bleeds to death over a period. The mice do not show bait-shyness, as they are unaware of the lethal effects and the poisons are safer for use in the

Small but effective

Mice have been known to eat some peculiar things, including curtains in Mayfair, organ keys in churches and a cannabis crop in Hampshire; more inconveniently, they have gnawed a hole in the side of a wooden hulled ship and chewed plastic beer pipes in the walls of a pub, with considerable loss of valuable liquid. German bank notes have been reduced to confetti – only redeemable as currency if the owner could piece them together again.

In 1981 the trains on the London to Cambridge line were stopped for three hours after mice had chewed through a 650 volt cable taking power to four level-crossings and three signals.

House mouse and trap.

presence of domestic animals, wild animals (but see Conservation) and humans as vitamin K acts as an antidote. These rodenticides are even used in the treatment of human blood disorders.

The house mouse shows much variability in its susceptibility to the poison; males are more susceptible than females and susceptibility increases with age in males but not females!

Warfarin is probably the most commonly known and commonly used anticoagulant rodenticide. However, resistance to this poison has been observed in house mice in Britain since 1960; it was also found in brown rats in 1958. The resistance occurs where warfarin has been used for long periods so that the mouse population becomes adapted to coping with the coagulation problems it causes. Some mice are better at coping with the poison and survive to rear young with the same genetic resistance to the anti-coagulant. The young will also be resistant and so a population of resistant mice builds up. These so-called 'supermice' are no different from ordinary mice in size or shape but simply have a slightly different genetic make-up.

The first occurrence of resistance was reported from Harrogate and it is now probably widespread in Britain. In order to control house mice other rodenticides had to be used and this led to the development of the 'second generation' anti-coagulants such as difenacoum, brodifacoum and bromadiolone. These poisons are used in the same way as warfarin and act in the same way but are more efficient and they are able to kill warfarin-resistant mice (and rats). However, there are populations of house mice that are cross-resistant, that is they are highly resistant to warfarin and also resistant to some of the second generation anti-coagulants. This has happened with bromadiolone and brodifacoum in

Birmingham and London and other remedies have to be used. Many other rodenticides are available and one example is alphachloralose, which is fast-acting. It acts by depressing brain activity and slowing the heart and respiration which cause fatal hypothermia. The substance is toxic to birds and so is recommended only for use indoors.

Other forms of resistance to poisons are also found in house mice. 'Behavioural resistance' refers to the behaviour of some mice which fail to take baits that are successful in other situations with apparently similar populations. Whether this change in behaviour is genetically controlled or simply the result of a changed environment where different foods are normally available we still have to find out. In the future we may have uncontrollable populations of mice!

Other weapons in the battle
If normal trapping and poisoning do not work, then fumigation (gassing),

chemical repellants and even chemosterilants have been tried. Ultrasonic sounds are supposed to repel rodents but tests show that it is difficult for these high-frequency sounds to penetrate areas where mice are found, as they do not travel far in air and may be absorbed by obstructions; it is doubtful if any ultrasound devices work efficiently.

The rodent-proofing of premises is another approach and even more ingenious is a new device that lets the occupier of the property know if mice have penetrated the building (though it doesn't actually help to remove them). This is done by placing special boxes with entrance holes in likely places for mice and detecting the mice by the use of an infra-red beam. When the beam is broken a signal is transmitted to a control box where a light and buzzer are activated.

Once caught in live traps, house mice can be removed humanely to another habitat, but make sure it is far away from your house – a mile ought to be sufficient (see Trapping small mammals).

Wild mice and men

Most of the small rodents are considered by someone to be 'pests'. The nature and severity of the damage they cause decides whether they are reckoned to be serious or not: for instance, species like the **harvest mouse,** which eats the ears of cereals, are not usually considered serious pests as the amount taken is so small. On the other hand, **wood mice** and the **voles** may all cause damage which is considered economically significant in agricultural or forestry terms, and control or preventative measures may have to be taken. The **house mouse,** of course, is a special case and has already been discussed (see previous chapter) and the possibility of disease transmission is considered elsewhere (see Parasites and disease).

Small mammals also become the unintentional targets for pest control measures aimed at invertebrates, **house mice** and rats in farmland, and are therefore conservation problems to some extent as well.

Mice and voles as pests

Short-tailed voles and their close relatives reach plague numbers and cause damage to agricultural crops in Europe. However, in Britain the **bank vole** and **short-tailed vole** are only serious pests of young trees in plantations, the latter being the most important, especially if there is a good growth of grass around the saplings, when they may ring-bark or completely eat the young tree at ground level. Damage to trees is most severe when population numbers are high or when there is snow cover protecting the voles' access. Many of the small mammals play a part in preventing regeneration in natural woodland as they will eat tree seeds, and **bank voles** in particular will eat seedlings of oak and beech and even small saplings of lime (see Food and feeding).

There are a number of ways of preventing the damage by **short-tailed voles** in forestry plantations. The easiest method is to keep the area around the stem weed-free by spraying with a herbicide or by hoeing; chemical repellants are also available. Another way is to protect the young tree with a tube of plastic which is often sunk into the ground (many different spiral expanding, split, rigid, square and round types are available). With these guards the tree gains protection from small mammals and, particularly with the rigid ones, protection from the cold, so that it grows much better in the first few years than an unprotected tree would. Experiments in forestry plantations suggest that the split tubes, without ventilation holes in them, sunk into the soil by 5 mm and with at least 200 mm height above ground give adequate protection from **short-tailed vole** damage.

The rigid tubes are often secured by cable-ties to a wooden stake and one of the disadvantages of this type of protection is that **wood mice** can chew their way into the tube where the cable-tie goes in (although the short-tailed voles can't climb as well and don't get in). The mice often make a nest at the base of the tree, chewing into the stem for food or to get nesting material!

Wood mice are not usually pests of commercial forestry (see above) but in sugar-beet fields they can cause significant damage by digging up and eating the pelleted seeds soon after they have been drilled in the spring. The seeds are spaced out a few centimetres apart in neat rows and it is not uncommon for a farmer to find a quarter of the sown seed taken from large parts of the affected fields; each pellet is removed by digging down in the soil and it is cracked open so that the seed embryo can be eaten. The clay coat of the pellet is usually impregnated with pesticides to prevent other problems but these pesticides, even if detrimental to mouse health, would not be ingested! It is common to see an abrupt line across a field where the damage has stopped and this may be the result of one mouse having found out about this new food supply when its neighbour has not.

There are a number of ways in which the farmer can try to combat the **wood mice** eating his sugar-beet seed without killing them. One is to re-drill the field later in the spring and hope that there is more natural food available when this is done. In addition, if there is rain soon after seed drilling the germination of the sugar beet may be rapid, leaving little time for the mice to find the new food source. Lastly, the mouse population is likely to be at a lower density later in the spring (see Numbers), again leading to less of a threat to the seeds.

If the farmer resorts to poison (see Of mice and men), it is best put under tiles, etc, to prevent birds eating it. Damage to the crop may be avoided if the drilling is left until late in the spring (see above), care is taken in placing the seeds at the correct depth of 25 mm (the closer to the surface, the easier they are for the mouse to dig up) and in not spilling any to 'give away' the fact that a new food source has arrived. The amount of damage varies considerably from year to year and has occurred since the early 1970s when the pelleted seed was introduced to Britain – the seed itself is very small and previously so much seed was sown that any losses were easily accommodated.

In gardens and outbuildings **wood mice** (and **yellow-necked mice**) may cause problems by digging up newly sown peas or eating flowers (e.g. crocus) or stored apples. They will come inside houses but are not as much of a pest as the **house mouse** because their feeding habits are more conservative (see Food and feeding and Of mice and men). If **wood mice**, or other small mammals, are captured with a live trap and released outside, make sure that they are taken a good distance from the house. **Wood mice** in particular are renowned for their 'homing abilities' and might come back without too much delay from distances of less than a mile!

A dead mouse is a dangerous mouse
It is not surprising that poisons put down to kill rats and mice will invariably kill other wild small mammals if they have access to them. This may happen in hedgerows and outhouses associated with farms and in urban situations if, for instance, poison is put down in a garden. These problems undoubtedly occur in isolated areas and any losses to the small mammal population would be quickly made up by others moving in. However, there is evidence to show that the poisons can be accumulated by predators of small mammals; this is thought to be one of the reasons for the decline in species like the barn owl which hunts in farmland and around farm buildings.

Just as disturbing, and more easily ignored, is the possibility of killing wild small mammals with agricultural chemicals not intended for this purpose. This probably does not happen very often nowadays, since the banning of many of the organochlorine pesticides which caused problems for wildlife in the 1950s and 60s, but it is still possible. The deaths of small mammals, if they occur, may go unnoticed because of the secretive nature of the affected species.

Farmers commonly use slug pellets on cereals and other crops when they are sown to help prevent slug damage during germination. The practice is even done routinely in areas prone to slug problems. One common slug-killing chemical is methiocarb, which is provided in cereal-based blue pellets (the colour helps to prevent birds taking them). Methiocarb pellets have been shown to kill **wood mice** which feed on them in laboratory tests and residues of the chemical may be found in animals living in the wild, but its effect on field-living mice is still uncertain. There is circumstantial evidence that after the slug pellets are spread on a newly drilled cereal field the mouse population shows a marked decline. However, the decline does not happen every time the pellets are spread and no direct evidence of dead **wood mice** is available.

The mice may not feed on the pellets every time they are available, just as sugar beet seeds are not taken every time they are drilled, and they may not take enough in the wild to cause any problems. The extra food provided by drilling the cereal seeds at the same time as the slug pellet applications may also divert the mice from eating the pellets. In addition, the pellets break up and distintegrate when they are wet and so heavy rain may prevent small mammals from eating them. The case against methiocarb is 'not proven'; at the moment there are so many **wood mice** living in non-cereal habitats that there are always many more to take their place if numbers do decline.

Large-scale field experiments comparing heavy pesticide use (including methiocarb and many herbicides, insecticides and fungicides which might have been needed in a 'bad' year) in cereal fields for five years found that the numbers of **wood mice** fluctuated markedly from year to year with spring lows and summer highs apparently unassociated with pesticide use, similar to changes in numbers in areas which had received many fewer pesticides. There is also evidence that if the pellets are drilled in with the seed, below the soil surface, the apparent decline in mouse numbers is not as great as when they are spread on the surface; this practice is sometimes carried out by farmers and might be kinder to mice if carried out as a general rule, at least until the subject is investigated further. There are also alternative slug poisons available such as metaldehyde which is commonly used as a slug pellet in gardens. Surprisingly, this chemical is used as much as methiocarb in some parts of mainland Europe and so the possible problems with methiocarb may be avoided by many of their farmers.

Conservation

It is unusual to think of small mammals as needing to be conserved, but the careful management of habitats is probably necessary for the continued existence of the island species (see Island mice and voles). The species found on the mainland may need to be conserved or encouraged in order to help conserve other species, especially predators like the barn owl or simply to maintain species diversity in new nature reserves.

Species such as the **harvest mouse** and **yellow-necked mouse** may seem to be uncommon but no special measures are needed as the former is well distributed in many Midland and southern tall vegetation/grassland habitats and the latter is locally but widely distributed over the same area. **Common dormice** have a similar limited and southerly distribution, but they are vulnerable to local extinction and conservation measures may be necessary in order to keep them at present levels. Happily, there is currently a fashion for coppicing woodland, and the older stages (15-20 years old) of long-rotation coppice are favoured by **dormice**. In areas where the woodland is suitable and **dormice** are absent it may be possible for introductions to be made to extend the distribution; these are currently being considered by English Nature.

Small mammals at the table

It is possible to encourage small mammals to feed on artificial surfaces just like birds at a bird table. The only problem is the prevention of the intervention of your friendly cat or rat while the mice and voles are eating!

The best position for a table is on a bank outside a window with one side up against the glass of a window and the other in thick grass/bushes, but tables further away at ground level will work

just as well. The idea is to make a 1 m square board (approximate size) safe from other visitors by putting a wire mesh (2.5 cm holes) cage over it about 60 cm high, supported on 2 × 3 cm struts. The wire is stapled to the struts and to the sides of the board and the front next to the window left open but attached to the window. The wire away from the house should be buried in the bank for 4–5 cm so that the mice and voles can get in only through the netting. In some circumstances it might be possible to join up the cage with a patch of cover in the garden by using plastic pipe (4-5 cm or more diameter) so that the animals can move between the cover and the feeding box.

The pipe connection is best used with a completely enclosed wooden box (except for the side facing the window) which may be as small as a shoe box; a small 'escape' hole (2-2.5 cm) should be made at one end of the box in case a weasel comes up the pipe when a mouse is feeding! The pipe should be as near to horizontal as possible so that the mice and voles do not slip – this means that the box will be close to the ground unless you have a steep bank nearby. The pipe should be 'baited' by pulling a rag covered in peanut butter through with a piece of string and it is best to have the pipe not more than 1-1.5 m long. With the netting-covered table it is useful to put moss on the floor and logs or branches so that the animals can feed under cover but still be seen from the window. The food for the small mammals can be placed on the table through the window or by making a flap opening in the wire which can be closed securely. Try watching at night with red lights (not noticed by mice and voles) and then with normal lights, which they ought to get used to. Observations at feeding points in the wild are also possible with red light if the viewer is very patient.

Sleeping bank vole.

How to help

In order to help the breeding and survival of predators such as the barn owl, their hunting habitats and small mammal prey need to be preserved. Small mammals may be encouraged by leaving wide field margins under hedgerows, allowing scrubby areas and other waste land to maintain a good grass cover which would favour **short-tailed voles**, and avoiding the use of poisons (and probably some slug pellets – see Wild mice and men).

If small mammals are to be encouraged in woodland, then variety is what is needed. Closed canopy woodland which shades out all the undergrowth, such as mature beech woodland and many coniferous woodlands, will probably have only **wood mice** throughout with possibly **bank voles** around the borders where the light allows the growth of ground vegetation. More open woodland and coppice, especially the younger stages with a good variety of ground cover species, will encourage most species of small mammals. Large-scale clearance of ground cover when trees are felled discourages **bank voles** but groups of brashings (sawn-off branches) will maintain the population, albeit at a lower level than that supported by normal abundant ground cover in an open canopy woodland. Planting trees with palatable fruits and/or leaves will encourage small mammals – particularly the masting species, like beech and oak which encourage periodic high densities.

A management practice to avoid is heavy grazing in both grassland and woodland. Grassland which is well grazed will not support many small mammals at all and grazed woodland will probably only support a low population of **wood mice** and possibly a few shrews.

Trapping small mammals

Live trapping of mice and voles is as much an art as a science. A trap mechanism can work very efficiently, but unless it is in the right place so that a mouse will come across it, then the exercise will be unsuccessful. Juveniles, before they have become fully independent, may move only a limited distance from the nest and traps set above ground will have difficulty catching these young animals even when they do venture out; their light weight will also make the springing of the capture mechanism difficult. Traps often rely on the inquisitive behaviour of the animal to enter the trap and this behaviour varies with individuals, some animals becoming 'trap-happy' and others 'trap-shy'.

The result is that trapping a sample of a small mammal population becomes only a sample of the 'trappable population' and may not be completely representative of the whole population. This is emphasized by the fact that pit-fall traps often catch juveniles and newly weaned animals much more efficiently than conventional mechanical traps. Further complications arise from the precise placing of the traps, as experienced trappers will often catch more animals than a novice. It is important to place the traps in regularly used runways which are easy to find in grassland but more difficult elsewhere. In woodland the best trap sites are against fallen branches or by the side of tree trunks and underneath some cover such as bracken or bramble; however, to obtain a proper sample open spaces should be trapped as well. It is common for small mammals to move along the sides of obstructions such as walls or branches on the ground and trap positions should reflect this habit if trapping is to be successful.

Types of trap

Most people are aware of the traditional mouse trap which is baited with cheese and kills the animal. This is still a very efficient way to control **house mice** in small areas and modified traps with large platforms for triggers (where the animal is caught simply by walking on the platform) are used to sample small mammals living in well-formed runways such as **short-tailed voles**. The traditional trap has to attract the mouse to the trigger with the bait of cheese or peanut butter but platform traps may be set without bait so that it is just the presence of an animal on the platform that results in capture.

In studies of small mammal populations it is usually necessary to catch the animals alive for examination and release them to follow their movements, survival, growth, etc. (This is perhaps now preferable for house pests anyway.) A live trap must be used and there are countless designs. In Britain the Longworth trap, which is made of aluminium, has been used to a great extent for this purpose. However, cheaper plastic traps (the Trip-trap) and simple alternative designs which can easily be made at home provide good alternatives for serious and casual studies. The important caveat, however, is that each trap design has its own characteristics and so catches may not be comparable if different designs are used.

How to set a trap

When traps are set they need to contain bedding such as hay or cotton wadding to keep the animal warm and cereal bait (oats or wheat are usually used) which is sprinkled outside and placed inside to attract the animal and keep it going. To comply with the Wildlife & Countryside Act (1981), live trappers of small rodents must take precautions to keep any shrews they catch alive and this is usually done by adding mince, maggots or fly pupae (fishermen's casters obtained from angling shops) to the food for mice and voles or providing an escape hole of 12-13 mm diameter to let out the shrews but not the larger mice and voles (see Keeping small mammals in captivity, p.120).

It should be pointed out that trapping is a serious business and should only be done with a justifiable purpose in mind. To be really sure of keeping every shrew alive the traps must be opened only during the day and visited every 2-3 hours.

Traps are usually placed in groups of two or more so that one capture does not prevent another; really the more traps at each point the better the sample, up to a limit. In practice the traps are usually placed in pairs in a line or grid formation. Traps are left overnight and visited at least every morning afternoon/evening. In line trapping, the area from which animals are attracted is unknown and although this is useful for general surveys and comparative studies a grid formation gives a more accurate idea of how many animals inhabit a particular area. However, there are still problems as animals normally living outside the edge of the grid may be attracted onto it to feed or explore, especially if neighbours are caught in traps, which would normally deter their use of the area. There are statistical ways to allow for this, however.

Trap lines or grids have different spacings according to the likely

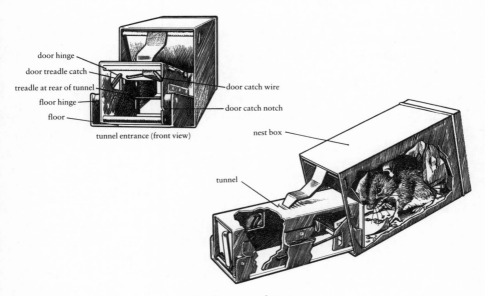

The Longworth trap.

movements of the population. For **short-tailed voles** and **harvest-mice** a 5-10 m spacing is common whereas for **bank voles** 10-15 m is suitable and for **wood mice** 15-20 m. **Wood mice** in arable land and **yellow-necked mice** might still be caught efficiently if a 20-25 m spacing is used. In most studies it is important to sample as much of the population on the grid area as possible and not miss animals that live in between trap points and do not move far enough to encounter a trap.

In other studies a sample from a large area may be needed; here a large spacing is necessary and more traps should be used at each point.

It is worth mentioning that bottles and cans left on the ground may catch small mammals which find they cannot get out and die inside; this commonly happens near lay-bys and places where litter is left. This is very unfortunate but the remains of the small mammals inside provide a further source of material for identification similar to that from owl pellets (see What the cat brought in and Truth in a tooth) and may be used for assessing species distribution and community composition in a particular habitat.

Another indirect method of sampling small mammals is to use sections of 3 cm plastic pipe with sticky tape stretched along the top of the inside, sticky side facing down (or pipes with sections sawn off and replaced by sticky tape). These tubes are placed in runways or other suitable places and the hairs of small mammals will be caught on the tape as they pass through and can be collected for identification (see Moulting and hair).

Marking small mammals

Simple marking methods may be used for short-term studies of small mammals. The easiest method is to clip off areas of fur with sharp scissors or just cut 'square' the hairs at the end of the tail in species like the **bank vole**. These marks allow individuals or groups of individuals to be identified and estimates to be made of population numbers from samples caught.

In practice it is common to catch a large proportion of the trappable population on a grid (see above) and so estimates may not be necessary. Small numbers of animals can still be marked with an individual fur clip so that movements, individual survival and behaviour towards the traps can be studied.

It is even possible to dust the fur with fluorescent paint dye in order to follow the movements of the animal. When released it will leave a trail of the dye on the ground or wherever it has explored, but this mark will not last for very long.

Long-term studies may require permanent marks and this should be thought about carefully as it may cause some discomfort to the animal. Early marking studies used numbered metal leg rings, similar to those used for birds, but these were found to restrict the growth of the leg in young animals and their use has been discontinued. Rings may be acceptable on adults, who have finished growing; but care should be taken to check that the leg is not being damaged by the procedure. Nowadays marking is easily done by purchasing serially numbered ear tags (see address list in Appendix). The tags are attached to the base of the ear and if this is done carefully and firmly the tag remains for the life of the animal and does not cause discomfort. Making a notch in the ear or punching a hole in it is permissible without a Home Office Licence because similar marks are made on domestic animals but this should be done only if really necessary. Other marks which take the end of a toe off (toe-clipping), preferably under local anaesthetic, provide many combinations of marks but because they may cause pain a Home Office Licence is necessary before this procedure can be carried out.

Trapping studies provide much useful information on the population ecology and habitat preferences of small mammals. The information in the chapter on Numbers is all based on these techniques.

Keeping small mammals in captivity

Mice and voles are easy to keep in captivity as long as their simple requirements are met. The species that are nocturnal, or mainly so, such as the wood mouse, yellow-necked mouse and house mouse, will need special arrangements such as being kept in a dark room with a red light during the day and a white light at night to alter their 'clocks' so that you can watch their behaviour without staying up all night. The harvest mouse, probably the house mouse, and the voles should be active for at least some time during daylight (see Mouse and vole territories).

Careful attention should be paid to the construction of a cage, as most mice and voles will try to escape from any hole large enough for them to squeeze through. Perspex or glass aquaria or home-made constructions with walls and floors of galvanized metal with a perspex front make suitable cages and perforated zinc or perspex with small air holes drilled in it may be used to seal the top. Wood and glass can also be used to good effect, especially for those species less well able to gnaw, such as the harvest mouse. Cages or aquaria should be at least 40 × 20 cm in their base and 20 cm high to allow adequate space for nesting and feeding, preferably 60 × 30 × 30 cm; the additional height will help to prevent mice jumping out when you remove the top, but all the mice, except the harvest mouse, will probably still jump out of these dimensions. Harvest mice might be better accommodated in a cage with a 30 × 30 cm base and a height of 45–60 cm so that tall stems and grasses can be placed in the cage and a more natural environment created. A small twiggy branch will give variety to any of the species' cages. Water bottles (obtainable from pet shops) can be supported on wires or a stand inside aquaria or held outside home-made cages so that only the drinking tube enters the cage. A 1–2 cm layer of sawdust on the floor is the best covering and as the animals always defecate in the same place it is advisable to clear this area frequently if not the whole cage. A nest box with cotton wadding or hay inside should be placed in the cage and this can be a simple wooden box or a jam jar. An activity wheel helps prevent obesity!

Food should be a cereal such as wheat or oats (or a balanced pelleted diet) for the mice and bank vole with a supplement of peanuts, a small carrot and quarter of an apple. Harvest mice will probably appreciate grass seeds and budgerigar or gerbil seed mixtures as well as earwigs, beetles and mealworms! The short-tailed vole will live on a cereal such as oats but with more vegetable matter (half an apple, a large carrot), and a daily supply of fresh grass; fresh turves may be used to cover the floor and provide food but they will need to be changed almost as frequently as sawdust and may obstruct observations. All species' diets may be varied by adding additional foods that might be eaten naturally, including some animal foods such as beetles and many insect larvae (see Food and feeding) on a trial and error basis; the food and water should be checked daily although they will last for longer if enough supplies are provided. There is no law against keeping shrews in captivity but you must have a licence if you intend to catch one in a trap (available from The Licensing Sections of English Nature, Northminster House, Peterborough, PE1 1UA, The Countryside Council for Wales, 43 The Parade, Roath, Cardiff, CF2 3ABZ, or Scottish Natural Heritage, 12 Hope Terrace, Edinburgh, EH9 2AS). A block licence may have been issued to your local Council to cover trapping by schools so it might be worth checking this first if you want to watch one in the classroom. It is also the case that if your cat brings a shrew into the house you are allowed to keep it to help it recuperate (see below). Note that the shrews should have 3–4 cm of floor covering and should be disturbed as little as possible. They feed on live invertebrates such as maggots or pupae (casters) obtained from angling shops; or meat, such as mince, from the butcher. The meat may go 'off' quickly and should be replaced and removed at least daily. Water shrews should not be given the opportunity to swim unless a burrow system is available in which they can groom and dry their fur.

It is safest to keep individuals singly (especially the shrews) but if two individuals are introduced to the cage at the same time, then they will usually get along, especially if two nesting jars are provided. It is not advisable to mix species except for the yellow-necked mouse and the wood mouse and even then it is advisable to check every day or so that they still like each other. You should never put new animals into the cage of another (they will probably fight) unless it has been thoroughly cleaned and they can start off on an equal footing. After a period of observation, or if the animals show signs of distress, you may prefer to return the animal to the wild; do this in a suitable place and note that obese animals might not be able to fend for themselves – it may be kinder to keep them! Breeding in captivity is possible for all the species.

If individuals are brought in by

the cat then it is possible to help them, especially if they are simply shocked and not otherwise injured. Keeping them away from the cat (!) warm in a tin or box with some cloth or vegetable bedding will often bring them round. When they are fully active again they can be released somewhere away from the cat's usual hunting ground.

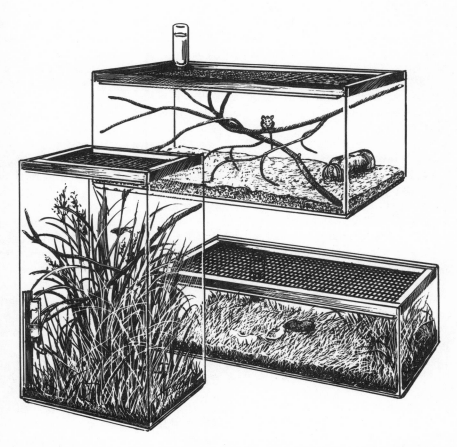

Captive environments for small mammals. In the background: aquarium for observing mice with 'minimum furniture'. In the foreground: more natural environments for keeping (left) harvest mice and (right) short-tailed voles. Food should be scattered to enable the mouse or vole to forage, and help minimize squabbles if more than one animal is being kept.

Shrews

The three mainland species of shrew have already been described (What the cat brought in) and their diagnostic features noted (Truth in a tooth). There are three other British species of shrew. The lesser white-toothed shrew (*Crocidura suaveolens*) occurs on the Channel islands of Jersey and Sark and all but some of the smaller Scilly Islands, and the greater white-toothed shrew (*Crocidura russula*) occurs on the islands of Guernsey, Alderney and Herm. The third species is the French shrew (*Sorex coronatus*) on Jersey which is very similar to the mainland common shrew in all but genetic composition and small differences in the skull.

As their names imply, the lesser white-toothed shrew is the smaller, being 3-7 g (⅙ oz) and 50-75 mm (2-3 in) in head and body length; the greater white-toothed shrew is 5-15 g (⅙-½ oz) and 60-90 mm (2.5.-3.75 in). Both have white enamel on their teeth and have prominent ears and both are greyish or reddish brown in colour. Their tails have short bristly hairs interspersed with long white hairs. Identification is not difficult as the two species do not occur together on the same island!

Both white-toothed shrews occur in France but the lesser white-toothed shrew has a southern European distribution and the greater white-toothed shrew a western European distribution going down to north Africa and encompassing all of Portugal, Spain, France and half of Germany; they were probably introduced to the islands by man. The lesser white-toothed shrew occurs in coastal and upper shore (boulder zone) habitats as well as in hedgerows, scrub, woodland and heathland, and the greater white-toothed shrew adds grassland and buildings to this list. They feed on a wide range of invertebrates, the lesser white-toothed shrew including crustaceans and larval flies from the sea shore. They are active mostly at night but, of the two, the greater white-toothed shrew shows more movement by day. The two species have a lower metabolic rate than the mainland shrews and can be relatively 'sluggish', eating only half their body weight in food each day. They breed from February/March to September/October and produce several litters each year. They are less territorial than the mainland species and will share a breeding nest in captivity.

The common shrew is the species most likely to be seen and heard on mainland Britain (it twitters as it explores), pushing through leaf litter and using the burrows of other small mammals. It feeds on a large number of invertebrate species including beetles, worms, slugs and insect larvae. The pygmy shrew occurs at lower densities than the common shrew and its feeding habits are similar but it does not eat worms or many large items. They need 80-90% (common shrew) and probably 120% (pygmy shrew) of their body weight in food each day. Breeding in both species takes place from April to September and up to four litters per female may be born each year in the common shrew. The sexes of both the common and pygmy shrew are distinguished by the inguinal bulges of the testes and in addition in the common shrew by the prominent lateral flank glands of the male; the female has 3 pairs of nipples (the water shrew has 5 pairs of nipples).

Both species are active during the day and night but the common shrew is less active during the day than the pygmy. They are vigorously territorial, aggressive (common shrews chase intruders and squeak loudly), and mainly solitary except when mating and lactating. The young maintain their own territories after weaning; they are more or less mutually exclusive and are possibly marked by scent from the flank glands in the common shrew. In the spring the males abandon their territories in search of females which are tolerant of them for about a day during oestrus. Home ranges (and presumably territories) are larger in the pygmy shrew than the common shrew and they increase in size in winter in the pygmy shrew, in contrast to the common shrew and small rodents.

The water shrew is adapted for swimming, with tufts of hair on its feet and a fringe on the underside of the tail; the hairs on the body are distinctly H-shaped, trapping more air than the common or pygmy shrew's hair and helping to repel water. The flank glands are prominent in adult males and fringed with white hairs. As with the other shrews they eat a wide variety of invertebrates as well as small fish and amphibians; usually freshwater crustacean species dominate and they need only 50% of their body weight of food each day. When they bite their prey they probably paralyse them with a venom from the salivary glands and after being bitten humans have a burning pain and reddening of the skin which lasts for a few days. Breeding and social organization and activity are similar to the common and pygmy shrews, although they are probably more tolerant of each other than the other mainland species.

Further reading

Bang, P., & P. Dahlstrom, *Collins Guide to Animal Tracks and Signs* (Collins, London, 1990)

Berry, R.J., *Biology of the House Mouse* (Symposia of the Zoological Society of London, No. 47, Academic Press, London, 1981)

Bright, P., & P. Morris, *The dormouse* (The Mammal Society, London, 1992)

Churchfield, S., *The Natural History of Shrews* (Christopher Helm, Bromley, Kent, 1990)

Corbet, G.B. & S. Harris, *The Handbook of British Mammals* (3rd Edition) (Blackwell Scientific Publications, Oxford, 1991)

Fairley, J., *An Irish Beast Book*, 2nd Edition (The Blackstaff Press, Belfast, 1984)

Flowerdew, J., *Woodmice and Yellow-necked Mice* (Anthony Nelson, Oswestry, 1984)

Flowerdew, J.R., J. Gurnell & J.H.W. Gipps, *The Ecology of Woodland Rodents. Bank Voles and Wood Mice.* (Symposia of the Zoological Society of London, No. 55: Oxford University Press, Oxford, 1985)

Gurnell, J., & J.R. Flowerdew, *Live Trapping Small Mammals. A practical guide* (Occasional Publication of the Mammal Society, No. 3. The Mammal Society, London, 1990)

King, C., *The Natural History of Weasels and Stoats* (Christopher Helm, Bromley, Kent, 1989)

Meehan, A.P., *Rats and Mice. Their Biology and Control* (Rentokil Limited, East Grinstead, 1984)

Teerink, B.J., *Hair of West European mammals: Atlas and Identification* (C.U.P., Cambridge, 1991)

Yalden, D.W. & P.A. Morris, *The Analysis of Owl Pellets* (Occasional Publication of the Mammal Society, No. 13. The Mammal Society, London, 1990)

Useful addresses

The Mammal Society, Conservation Office, c/o The Department of Zoology, University of Bristol, Woodland Road, Bristol, BS8 1UG (For their publications, membership and information on mammals)

Penlon Ltd, Radley Road, Abingdon, Oxon, OX4 3PH (Suppliers of Longworth traps)

North-West Plastics Ltd., Worsley, Manchester, M28 4AJ (Suppliers of 'The Living Trip-trap')

Address for ear tags: Prof. E. Le Boulenge, Unité Biométrie Université Catholique de Louvain, Place Croix du Sud, B-1348 Louvain-la-Neuve, Belgium

Index

If you have enjoyed this book, you might be interested to know about other titles in our **British Natural History** series:

BADGERS
by Michael Clark
with illustrations by the author

BATS
by Phil Richardson
with illustrations by Guy Troughton

DEER
by Norma Chapman
with illustrations by Diana Brown

EAGLES
by John A. Love
with illustrations by the author

FALCONS
by Andrew Village
with illustrations by Darren Rees

FROGS AND TOADS
by Trevor Beebee
with illustrations by Guy Troughton

GARDEN CREEPY-CRAWLIES
by Michael Chinery
with illustrations by Guy Troughton

HEDGEHOGS
by Pat Morris
with illustrations by Guy Troughton

OTTERS
by Paul Chanin
with illustrations by Guy Troughton

OWLS
by Chris Mead
with illustrations by Guy Troughton

POND LIFE
by Trevor Beebee
with illustrations by Phil Egerton

PUFFINS
by Kenny Taylor
with illustrations by John Cox

RABBITS AND HARES
by Anne McBride
with illustrations by Guy Troughton

ROBINS
by Chris Mead
with illustrations by Kevin Baker

SEALS
by Sheila Anderson
with illustrations by Guy Troughton

SNAKES AND LIZARDS
by Tom Langton
with illustrations by Denys Ovenden

SPIDERS
by Michael Chinery
with illustrations by Sophie Allington

SQUIRRELS
by Jessica Holm
with illustrations by Guy Troughton

STOATS AND WEASELS
by Paddy Sleeman
with illustrations by Guy Troughton

URBAN FOXES
by Stephen Harris
with illustrations by Guy Troughton

WHALES
by Peter Evans
with illustrations by Euan Dunn

WILDCATS
by Mike Tomkies
with illustrations by Denys Ovenden

Each title is priced at £7.99 at time of going to press. If you wish to order a copy or copies, please send a cheque, adding £1 for post and packing, to Whittet Books Ltd, 18 Anley Road, London W14 0BY. For a free catalogue, send s.a.e. to this address.